to
Linda and John
and
Michael and Stephen Fitzpatrick
with affection

Contents

CHAPTER   5   ELECTRICAL COMPONENTS PART II   137

Contents   **xiii**

Preface

This book should be used in conjunction with the 1996 edition of the National Electrical Code® (NEC). The purpose of the Code is to ensure that all electrical wiring is installed correctly with "correctly" meaning with a view toward maximum safety for both person and property. The National Electrical Code is a reference work and should be consulted if there are ever any questions about electrical work. Unlike the Code, this book supplies relevant data on electrical installation following the specific guidelines supplied by the 1996 NEC. In an effort to help the reader, references to particular articles are supplied and the required page numbers in the 1996 NEC are given. By following the instructions in this book and consulting the 1996 NEC as well, you have an assurance that the work will most likely pass inspection by municipal electrical authorities.

NOTICE TO ALL READERS OF THIS BOOK

The use of electricity in the home is increasing steadily, and so is the accompanying risk of being hurt by it. References are made throughout this book to the need for safe practices and for the use of devices that protect against the possibility of leakage currents and their danger. The reader is also cautioned always to disconnect power before doing any electrical work and explanations are supplied on how to do this.

Neither the publisher, Prentice Hall, nor the authors, Martin Clifford and Jerrold Clifford, can assume any responsibility for any misinterpretations of the material supplied in this text or for any errors made by the reader. Readers are urged not to undertake any wiring that is beyond their capabilities and to consult and use the services or the guidance of licensed professional electricians in the event of any doubt or need for further information.

Every home is somewhat electrically inadequate; this applies to both older homes and those that have just been constructed. This is due to the growth in use of both electrical and electronic appliances and can be recognized by the extensive use of cube taps and strip outlets. The results are increasingly poor operating characteristics and, in a worst-case situation, an electrical fire. In some instances the electrical wiring and receptacles are simply inconvenient or inadequate or both. Electrical power can be used properly, but this requires an understanding of electricity and the changes that can be made, an awareness of new electrical components, and knowledge of the devices that contribute to electrical safety.

Chapter 1 begins with some comments about the 1996 National Electrical Code and calls your attention to the fact that the electrical inspection departments of various municipalities use the NEC as a guide, thus giving that book the effect of legal authority.

No electrical installation can be any better than the components that are used. It is always advisable to use products that carry an Underwriters' Laboratories logo, for that is your assurance that they have passed rigorous testing. Chapter 1 is also your introduction to electrical safety, both indoors and out.

All electrical alterations require an understanding of the language of electricity and its three basic units: voltage, current, and resistance. It is essential to avoid "cookbook" electrical changes and repairs, that is, to make changes and modifications in an existing electrical installation without understanding the basics. These are described and explained in depth in Chapter 2, Basic Data for Electricians.

Electrical power is delivered to the home by electric utilities. Their responsibility ends at the entrance to the service box. Beyond that point, the power is distributed by electrical branches. These can be rerouted, extended, or augmented. All branches must use wire of an adequate size, correct current-handling circuit breakers or fuses, and wire that follows the rules of the 1996 NEC guidelines. Chapter 3 is a detailed explanation of wire, ampacity, wire types, insulation, and solid and stranded wire. Sometimes wires must be joined, and this chapter supplies a working description of the various kinds of splices, joints, and taps.

Chapter 4 contains a detailed description of fuses, circuit breakers, plugs, and receptacles. Circuit breakers and fuses are important because they offer protection against excessive current demands or short circuits. These components are important insurance against uncontrolled current flow. A fairly recent development is the ground-fault circuit interrupter, a device that adds further protection. This chapter also supplies detailed data about plugs and receptacles. These components permit the easy connection of electric devices to branch power lines.

The flow of electric current must be controlled, and while there are various components that can do this, possibly the most commonly used is the switch. Switches can control single loads or two or more loads from a single location. By using switches it is possible to direct the flow of current in either one or more directions. Chapter 5 gives a detailed description of a large number of switches and also the enclosures, known as electric boxes, into which they are placed. Even a component as simple as an electric box, with no moving parts, must be carefully selected and properly installed.

With the preceding five chapters as a guide, Chapter 6 is an introduction to practical wiring. This chapter explains how to connect a ceiling, attic, or basement light and how to tap in on a switching circuit. Not all wiring is easy, since it can involve running wires behind existing walls. It is often possible to take advantage of branch wiring installed behind the walls during the construction stage.

Chapter 7 is actually a continuation of the preceding chapter, for it emphasizes various wiring installations, including the installation of dimmers, how to fish wire, how to connect a ceiling fan, and how to upgrade receptacles.

Electrical wiring involves more than just working with wires and lighting fixtures. Chapter 8 explains electric components such as transformers, capacitors, inductors (coils), and relays. Like the preceding chapters, this chapter also has practical information. Transformers are used for door chimes and buzzers, capacitors are required for some types of motors, and relays can be used to operate electrical circuits at a distance.

Electric lighting is used both indoors and outdoors. Chapter 9 provides a detailed description of incandescent tungsten light bulbs, fluorescent lights, halogen lights, and some of the most recently developed bulbs. These offer improved lighting and operating economy. Electric lights are not only essential for indoor illumination, but have also become extremely important for outdoor lighting.

For many people, motors are mysterious components, but actually they are simple electrical units. A motor is a transponder that changes electrical energy to its mechanical equivalent. There are numerous types, ranging from the fractional-horsepower clock to those used for operating washing machines. As is the case with other electrical devices, motors are available in a large variety of sizes and operating powers. Direct-current (dc) motors are detailed in Chapter 10.

Chapter 11 is a continuation of Chapter 10 but discusses alternating-current (ac) motors exclusively. This chapter contains a description of the various types of ac motors, torque (or turning power), centrifugal switches, motor speed control, and motor rotational control, plus a discussion of generators. Three-phase power is also covered.

Chapter 12 supplies a large number of solved problems; we detail the problem, indicate the formulas to be used, and explain the techniques for arriving at a solution. The problems cover the same material discussed in the earlier chapters. This chapter should be of particular importance to those who expect to take examinations.

For a long time, electricity and electronics were regarded as subjects that had no relationship. However, the two subjects are beginning to join forces, and in time electricians and electronic technicians and engineers will need to learn both subjects. Electronics is now being used to control stepper motors and motor direction and for automatic starting and stopping of motor shaft rotation.

Martin Clifford
Jerrold R. Clifford

Note: Throughout this book the use of the words NEC or Code shall be understood to be a reference to the National Electrical Code.

Acknowledgements

We would like to acknowledge with thanks and appreciation the contribution of electrical data, technical information, and drawings in the literature and catalogs generously supplied by these companies.

Brooks Electronics Inc.
Edison Lighting Co.
Electripak Consumer Products.
Florida Power & Light Co.
Franzus Co.
Fyrnetics, Inc.
Gem Electric Manufacturing Co., Inc.
General Electric Co., Wiring Devices
Gilbert Manufacturing Co., Inc.
Gould Electronics
Hubbell Incorporated, Wiring Device and Lighting Divisions
Jersey Central Power & Light Company
Leviton Manufacturing Co.
Mulberry Products, Inc.
Penton/PC, Inc.

Radio Shack, Division of the Tandy Corp.
Sears
S. L. Waber, Inc.
Small Motor Manufacturers Association
Wiremold Company

Chapter 1

Electrical Overview[1]

Electrical work is different in one respect from activities such as carpentry, plumbing, and furniture refinishing. Electricity involves a greater element of risk. Improperly used, it can result in a fire or an electrical shock. Wrongfully used, it can result in an increase in operating expenses. Often, the difference between electricity as a servant or a master is paying attention to details. The decision as to which it will be is not made by electricity. That decision can only be made by humans.

Basically, there are two types of electricians: amateur and professional. The amateur electrician has various levels of expertise.

Professional electricians can be classified as trainees, novices, apprentices, expediters, journeymen, and master electricians. For commercial installations, an electrician in charge of maintenance of supplies or control over work flow is an expediter. The difference between an apprentice and a journeyman is in the amount of experience and electrical knowledge. Both work under the supervision of a master electrician, someone who is licensed by his or her municipality or county to do electrical work. In some areas, journeymen electricians are also licensed. In that case the responsibilities of each class, master and journeymen, are separately defined.

[1](*NEC®: Definitions, Article 100; page 21*).

MUNICIPAL OR COUNTY ELECTRICAL CODES

The laws governing electrical changes or completely new installations vary in different municipalities or counties, and while they generally follow the guidelines set up by NEC, there may be special requirements because of local conditions. In any event, it is always advisable to check with local authorities to determine the extent to which electrical work can be done without professional supervision. In some instances the municipal authority will call attention to the fact that it issues competency licenses and certificates (Figure 1–1). The municipality or county may supply a notification form indicating that certain electrical work is to be done, with that notice posted on site. Various areas may have an electrical inspector check on work that is done prior to or subsequent to its completion, or both.

Electric utilities work closely with local authorities and, as a general rule, will not arrange to have power turned on until they are satisfied that an inspection certificate has been obtained.

A municipal or county electrical code is not simply a request but has the force of law. For the most part, this code is evidence that, at the time a new home or commercial property is being constructed, or an older home is being updated, or when an electrician's services are to be used for electrical modifications, legal requirements have been satisfied.

There are possibly areas with no local licensing or inspection requirements, such as rural or farm locations. These may be governed by state laws.

In all cases it is best to make inquiries to ensure full compliance. Failure to do so and the inability to produce a permit can result in a fine or, in the event of a fire that can be attributed to an electrical installation, can mean not only a fine but also the loss of insurance coverage, even though all premiums have been paid. Be sure to read your fire insurance policy or consult your insurance company or their broker representatives prior to any electrical changes.

There is no nationwide law governing electrical installations. A permit may

---

Special Notice

A NOTICE–Broward County requires competency licenses for individuals involved in construction & home improvement services. Be advised to check on competency licenses & certificates. Broward County: 785-4655. To register a complaint: Broward County Consumer Affairs Division: 357-6030

Figure 1–1. County notice of requirement for competency license.

or may not be required. Electrical inspection following installation may or may not be available. Information as to what may or may not be done can be supplied by a licensed master electrician, municipal or state authorities, or insurance brokers. Such an inquiry is always the first step prior to making electrical modifications. If the property is a leased apartment, office, or factory space, it is not the property of the renter. No electrical changes can be made without prior consent of the owner. Subsequent electrical modifications are invariably considered as belonging to the owner of the property.

THE NATIONAL ELECTRICAL CODE®

The National Electrical Code, abbreviated as NEC, is widely used, not only in the United States, but worldwide. The basic goal of NEC is to achieve an electrical fire-free environment. Indirectly, the code has the force of law since electrical inspectors and master electricians use the NEC as guidelines.

The function of the National Electrical Code is to supply guidelines for the installation of electrical components, wiring methods, and the kinds of component materials, with an emphasis on safety in their operation and functioning: safety from fire and protection against electrical shock to installers and users. The code is not concerned with efficiency or convenience or with the possible future expansion of an electrical system. This does not mean that these are of no consequence. They are important, but as far as NEC is concerned, they are considered only from the viewpoint of safety.

The rules set up by NEC are not arbitrary or capricious, but are based on the extensive experience of large groups of electrical specialists. Ordinarily, for commercial buildings an electrical installation or the modification of existing wiring is done by an electrical contractor. A homeowner may also make electrical changes. However, municipal authorities must be notified and the work must be subject to their inspection. NEC is not concerned with whether an electrical installation is done by an amateur or a professional provided, it is done following recommended procedures. It is not NEC's function to see to it that you get a certificate of compliance from your municipal or electrical authorities. That is the responsibility of the owner of the residence, commercial building, store, factory, or any other structure where electricity is used. When electrical work is done by a subcontractor or an independent master electrician, it is his or her responsibility to obtain a certificate of compliance.

Since a typical homeowner may not be familiar with NEC guidelines, an advisable method is either to make use of the services of a licensed master electrician or to do the work under his or her supervision. Even if a copy of the latest

NEC volume is available, local conditions may require special attention and may not be covered by NEC rules.

The National Electrical Code is published by the National Fire Protection Association (NFPA), Batterymarch Park, Quincy, MA, 02269-9101. The code is revised every three years and it is advisable to work with the most recent issue. While a copy of the Code can be obtained from a local library, it may be out of date. The safest procedure is to buy a copy directly from the NFPA, or, if working with an electrician, to make sure he or she is familiar with the latest NFPA provisions.

It is not the intent of the National Electrical Code to supply electrical design information. That is done by electrical design architects, master electricians, or electrical contractors.

THE ELECTRICAL PERMIT

Before beginning an electrical wiring project, consult your local electrical authority. One advantage of working with licensed electricians is that they are familiar with existing rules and regulations. There may be a municipal fee for the inspection of an electrical installation.

Use of Materials

For those not familiar with electrical components, the number and variety of parts can be extremely confusing. It is often advisable to allow your electrician to make the selection. However, if you wish to do so yourself, make a list of the parts and include the electrical characteristics of each. Alternatively, keep the parts containers since the data on the parts may be listed on them.

If your home or apartment is covered by fire insurance, and it should be, discuss contemplated electrical changes with the company or their agent. Your fire insurance policy covers your home as it existed at the time the contract was signed. That coverage is not automatically transferred to subsequent electrical changes.

Old Homes versus New Homes

Even a brand-new home may require electrical changes. New receptacles may be needed. The current-carrying capacity of an electrical branch may be inadequate. New fixtures may be required. Before making a purchase, use the services of an electrician to learn just what changes need to be made and add the cost to the purchase price.

These comments also apply to an older home, but much more so. All or part of the existing wiring may need to be replaced. You may want to replace a fuse box with a box that uses circuit breakers, for these are much more convenient. You may want to have additional receptacles installed. It is much easier to make electrical changes at the time the house is vacant so that you move into an electrically safe and up-to-date home. Again, add the cost of electrical changes to the purchase price of the home.

Four factors are involved in any electrical installation.

1. It should follow NEC guidelines.
2. The components to be used should carry the UL label.
3. The installation or modification of the electrical system should have the approval of a local or county electrical authority.
4. The user of the system must not do anything that will negate any of the first three factors listed here.

UNDERWRITERS' LABORATORIES®

It is possible to follow NEC guidelines and also to obtain a municipal certificate of approval and yet to have an electrical installation that is a fire hazard. Neither NEC nor a municipal code specifies the quality of electrical components that are used. These can be substandard and are sold by electrical parts manufacturers using low prices as the sole criterion.

To make sure that the electrical parts they manufacture meet minimum standards, the parts are submitted to the Underwriters' Laboratories, also known as UL®, for test and evaluation. The selling price of these parts is not a consideration; hence price approval by the UL is not a factor. The UL was set up for making detailed, exhaustive tests of electrical components such as switches, fuses, relays, circuit breakers, and wire. If a component is modified in some way, then it must be resubmitted for UL approval. If an electrical part passes all its tests successfully, it is then included in the UL's *List of Inspected Electrical Components.*

UL approval is not only applicable to electrical parts but to appliances as well. In the case of an appliance UL approval indicates that its constituent parts have also been UL tested.

If items meet all UL tests, they may then carry a label indicating UL approval. Various types of labels, a few of which appear in Figure 1–2, can be used. The UL label can be a label adhering to the surface of a product. In the case of wire, it may be printed on the outer insulation. It can also be in the form of a tiny metallic disc.

Manufacturers of electrical products are interested in UL approval since it

Figure 1–2.   A few of the various UL labels. Some of these carry physical, electrical, or descriptive data.

gives them a competitive edge. This does not mean that a product has UL approval in perpetuity. Those that go through material or design changes must be resubmitted. The UL stamp or label means the electrical product has met tests imposed by an outside agency.

A UL listing is not a guaranty or warranty. The responsibility for the nonfunctioning of an electrical part within the guaranty or warranty period is the manufacturer's. Electrical parts should always be operated within the limits of their specifications, as indicated by the manufacturer. Thus, if a component carrying a UL logo is specified by the manufacturer as having a maximum current capacity of 2 amperes, it may not be used in a 4-ampere circuit. One function of the Underwriters' Laboratories is to make certain that electrical products do meet such specifications claimed by the manufacturer.

LINE VOLTAGE AND LINE FREQUENCY

Electric utilities supply alternating current (ac) voltage known as line voltage to homes, offices, farms, and industry. This voltage can have two values, 120 and 240 volts, but these are nominal (*NEC: Article 100; page 24*). There is no standard nor can there be, since the actual amount of voltage available at a receptacle depends on the voltage output of the generator used by the utility, its distance to electrical receptacles (outlets), and the amount of load. On very hot and humid days, with strong use of air-conditioning systems, the amount of available voltage can drop substantially below average. This loss of voltage, called a brownout, can result in poor operation of electrical appliances. In severe instances, the voltage drop will be a blackout, with complete inoperation: no lights, no elevator service, no radio, no television.

The voltage marked on an electrical appliance indicates one possible voltage. The actual voltage at an electrical receptacle may be higher or lower. Some appliances are marked 110 volts, with others indicating voltages up to as much as 125. As in the case of electrical utilities, there is no standardization. The fact that an electric light bulb is marked 121 volts does not mean a utility will supply this amount.

While voltage can be a variable amount, the frequency of that voltage, measured in hertz (Hz) or cycles per second, is constant. It can vary slightly, possibly from 59 to 61 Hz, but over a 24-hour period its average will be 60 Hz. For this reason, electric clocks operating from a receptacle are remarkably accurate.

DELIVERY OF ELECTRICAL POWER[2]

Electrical power can be supplied by a utility via a two-wire line having a potential of 2400 volts. This is supplied to a voltage step-down transformer having a primary winding and a two-section secondary (Figure 1–3). This transformer has two advantages. It has a voltage step-down whose turns ratio is 2400 to 240 or 10 to 1. The available secondary current is stepped up by the same amount. The twin secondary winding supplies a pair of 120-volt lines or a 240-volt line, depending on how the voltage is measured. The voltage is 120 volts with respect to a grounded neutral. It is also 240 volts when measured from one high tension line to the other. As a consequence, this arrangement supplies both 120 volts and 240 volts. The 240-volt line operates high-current devices, such as an electric stove, clothes washer, or dryer. The 120-volt line is used for lower-current devices, such as TV sets, radio receivers, electric fans, and lights.

Note in Figure 1–3 that the neutral (*NEC: Article 100A; page 29*) wire is grounded. Actually, two wires are used for ground. One of these wires is known as the *neutral;* the other is referred to as *ground.* At one time just a ground wire was used, but this was unsatisfactory since it was dependent on conductors that could either be interrupted or could have an unwanted amount of resistance. The ground wire and the neutral are connected and so they can be regarded as being in

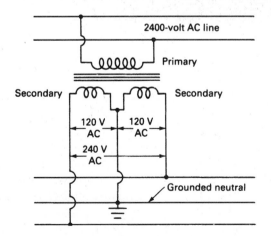

Figure 1–3. Connections to a power-line transformer.

[2](*NEC: Article 250–5; page 123*)

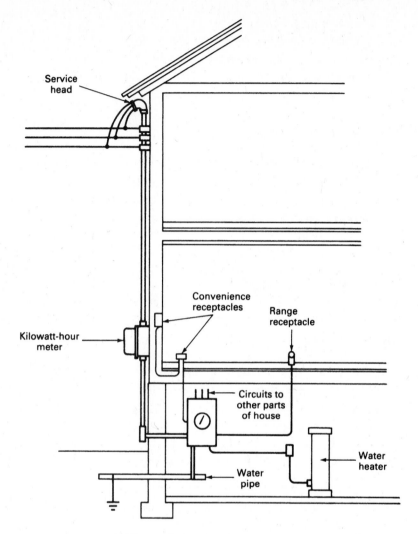

Figure 1–4. Service entrance.

parallel. Because of this connection, the overall resistance of the neutral and ground is lower than it would otherwise be.

No current is ever lost and no current is ever "used up." The total amount of current supplied by the generators of a utility to a home, farm, or any establishment using electrical power must flow back to those generators. Since the current is *alternating,* the term as used here means that the current circulates back and forth between the utility and the load. The load consists of all the electric devices at work in a home, or all the electrically operated lights, air conditioning, and machines in a factory or office building.

A local utility supplies electrical power by means of wires having a low gauge number. Following the step-down transformers there are three such wires, consisting of two hot lines and one neutral. The neutral is the reference, with each of the two other wires having a potential of 120 volts compared to the reference. The voltage measured from one hot lead to the other is 240 volts.

The three wires from the utility's power lines are led into a *service head* (Figure 1–4) mounted at the top of a long length of conduit. These wires are not straight and taut, but form loops that hang downward. Known as *drip loops* (*NEC: Articles 230–52; page 97*) the wires are arranged in this manner to keep water from entering the service head. The wires out of the service head must extend for at least 3 feet before connecting to the utility company's power lines. Whether the wires are arranged horizontally or vertically, the center wire will probably be the neutral.

The three wires continue on to a kilowatt-hour meter for the measurement of electrical power usage. In apartments and older homes, the meter is indoors, but the present method for homes or for stores is to mount the meter (or meters) outside. The advantage is that the utility's meter reader has access to the meter at all times.

From the meter the three wires enter a metal enclosure referred to as a *service box* or sometimes called a fuse box (*NEC: Article 230; page 96*) or a circuit breaker box. The two hot leads pass through a pair of main fuses or circuit breakers, but immediately preceding these there is a switch capable of opening both hot leads simultaneously (Figure 1–5), but not the neutral or ground wire. This wire is continuous (*NEC: Articles 300–13; page 167*) is never interrupted, and makes a direct connection to the incoming neutral. On older-type fuse boxes, the switch is controlled by a handle directly on the outside of the box.

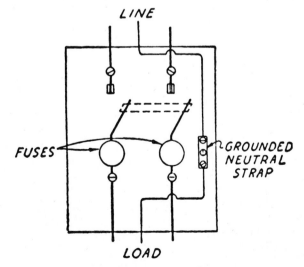

Figure 1–5. Switching arrangement of a fuse box. Note that the neutral wire is not fused. The two main fuses must carry the total current.

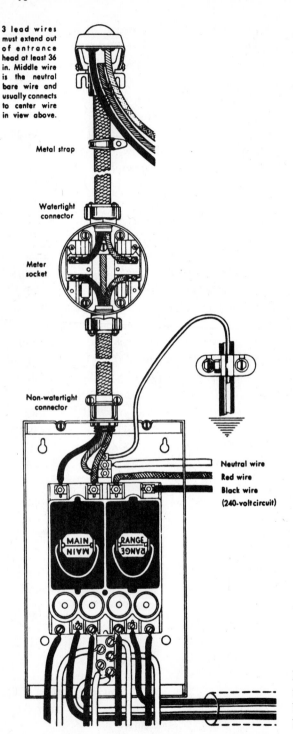

3 lead wires must extend out of entrance head at least 36 in. Middle wire is the neutral bare wire and usually connects to center wire in view above.

Metal strap

Watertight connector

Meter socket

Non-watertight connector

Neutral wire
Red wire
Black wire
(240-volt circuit)

MAIN

RANGE

Figure 1–6. Power line entrance to a service box. The beginnings of branch circuits are formed in this box.

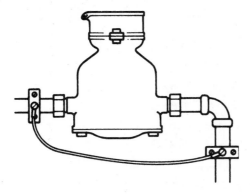

Figure 1–7. Jumper cable across water meter.

A circuit-breaker box has no outside handle. Instead there will be an in-box control identified as the main circuit breaker. This performs the same function as the outside switch on the fuse box. Keep the service box closed except when changing fuses or resetting circuit breakers.

Figure 1–6 shows a switching arrangement in a fuse box. The entry wires may be color coded red, black, and white. The red and black wires are the hot leads; the white wire is the neutral. The neutral will be fastened to a ground connector on the box, and from that point the neutral, usually a heavy, bare, stranded cable, will continue on to a section of a water pipe that precedes the water meter. A jumper cable (Figure 1–7) across the water meter helps maintain the continuity of the neutral. (*NEC: Article 250–71a; page 139*).

Branch Formation[3]

Figure 1–8 shows the formation of branch circuits used for supplying electrical power throughout. Each branch consists of a pair of power lines. A pair of such lines, with one connected to the neutral and the other to either a black lead or a red lead, supplies 120 volts. A pair connected to the red and black leads supplies 240 volts.

Some branch lines may be rated at 10 amperes and others at 15, 20, or 30 amperes. Only the hot leads are fused or protected by circuit breakers. The neutral lead, the ground lead, and a wire connected to either of the black or red leads forms a branch. The number of branches depends on the original design by the electrical contractor. The branches connected to the service box may be housed in either rigid or thin-walled conduit or nonenclosed indoor-type, plastic-sheathed cable, sometimes referred to as *Romex.*

The circuits or branches are connected to the main power lines. The hot leads of circuits 1 and 3 are connected to the red wire and to the neutral (cold

[3]*(NEC: Article 210; page 52; Article 220; page 68)*

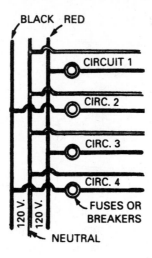

Figure 1–8. Formation of branch circuits.

lead), circuits 2 and 4 to the black wire and the neutral. All the branches form 120- or 240-volt power lines that go to the various receptacles.

Do not confuse the words *hot* and *cold* with temperature or with positive and negative. Either the hot lead or the cold lead can be alternately positive or negative, and the number of times this happens depends on the frequency of the applied voltage. A hot lead always remains a hot lead, and a cold lead always remains a cold lead. The concept of changing positive and negative applies only to ac. If the voltage source is dc and that source is not transposed, a positive lead remains positive, and a negative lead does not change its polarity either.

INTERRUPTION OF ELECTRIC SERVICE

If electric power is interrupted, turn off any appliances that will go on automatically when service is restored. This includes furnaces, air-conditioners, electric water heaters, refrigerators, freezers, and pumps. Also make sure electric space heaters, washing machines, clothes dryers, dishwashers, and TV sets are off. The reason for this is that, when service is restored and a number of appliances come on at once, the circuits may be overloaded. Leave a single lamp turned on as a signal to indicate that service has been restored.

If, at the time of electric power interruption, a device such as a washing machine is partly through its function, turn its dial to the off position.

Every home should be equipped with a flashlight to use when service is discontinued. Check the flashlight at least once a month to make sure it is functioning. Keep a spare set of batteries adjacent to the flashlight as a backup.

Downed Electric Power Lines

During a storm, electric power lines may come down. Treat all such wires as if they were live. Do not touch or try to move them. Keep children away. Also, remember that many things, such as metal fences, aluminum siding, and water, are electrical conductors. Report all downed wires to your local power utility. If their phone line is busy, call the police.

Power Outages

While power outages are often associated with stormy weather, they are possible under other conditions. A tree may fall against power lines, a strong wind may damage them, or there may be a system problem in the utility.

An outage may be determined by its characteristics. There may be a drop in the available voltages, leading to what is called a *brownout,* or a complete absence of voltage, leading to a *blackout.*

Preparation. In the event of either type of power loss, it is well to be prepared. Set up an outage kit containing a list of emergency phone numbers, a flashlight in working order, candles and matches, bottled water, extra blankets or sleeping bags, canned foods, a mechanical (nonelectric) can opener, and a battery-operated radio. The outage kit should be in an easily accessible and well-remembered location. Keep a list of emergency numbers in the kit: police, fire department, local utility, and those of friends, relatives, and neighbors.

Weather outage. A weather outage may last a short or long time, depending on its cause and the ability of the utility to reach the trouble spot (or spots). Under some weather conditions, it may be possible to predict an outage. Turn the controls of the refrigerator to its coldest position. Food will stay frozen between 36 and 48 hours in a fully stocked freezer if the door is kept closed. Food in a half-full freezer will keep 24 hours. During the winter, food can be stored in a cold area outside the home. Use caution. Be careful about using uncooked meat and fish or any foodstuff containing cream.

Report the outage to your local utility even though you may think they must be aware of it. Doing so will help them pinpoint the extent of the outage.

Turn off all appliances, particularly major ones such as air conditioners, the electric range, and the clothes dryer. A single light left on will signal when power has been restored.

If you try to cook indoors during an outage, remember that the biggest hazards are fire and carbon monoxide. Never burn charcoal indoors because it releases carbon monoxide, a potentially fatal gas. Use a fireplace, a Sterno or its equivalent if you must cook.

If any member of the household is dependent on electrically operated medical equipment, such as a respirator or a kidney dialysis machine, notify the customer service center or business office of the electric company at once. They will try to restore power to life-support equipment customers on a first-priority basis. Before any outage occurs, discuss the medical problem with the utility. No utility can guarantee continuity of service, so it is important that medical equipment have a battery backup. Check to make sure that the battery is fully charged at all times.

If your service box is fuse operated, keep a supply nearby. To replace a blown fuse, shut off the main switch and replace the burned-out fuse with a new one of the correct current rating. It will be helpful to have an extra flashlight (other than the one in the outage box) near the service box. If the service box is equipped with circuit breakers, locate the open breaker and reset it.

Rolling blackouts. On some rare occasions when extreme hot or cold temperatures create new peaks in demands for electrical energy over more than just a few hours, a utility may have to initiate a rolling blackout. These controlled power interruptions usually last no longer than 15 or 30 minutes and prevent the area from experiencing an uncontrolled blackout. In such severe weather emergencies, the utility will issue radio and television weather and power updates.

Rolling blackouts are an emergency measure and are used only as a last resort. When such a blackout occurs in extremely hot weather conditions:

Set the air-conditioning thermostat no lower than 78 or turn it off.

Close drapes to keep out the sun.

Turn off all unnecessary lights.

Keep the refrigerator door closed as much as possible.

In general:

Refrain from nonessential uses of electricity until the emergency has passed.

Turn off all devices that may have an associated motor, such as a pool pump.

Unplug appliances.

Turn appliances on one at a time when power has been restored.

Telephone your utility only if power has not been restored within 30 minutes and other homes in the area have power.

Use a GFCI.[4]   Use a ground-fault circuit interrupter (GFCI) where there is easy access to water: in bathrooms, basements equipped with a sink, kitchens, garages having a water line, and laundry rooms. These units will trip under short-circuit conditions and protect the appliance user. GFCIs are becoming common-place for building code compliance in new structures and can be easily installed in older homes.

Electrical Safety Outdoors

Electrical safety procedures are as essential outdoors as indoors. If you decorate outdoors, look up! Don't raise ladders or other extended objects into power lines. Don't assume because a ladder is made of wood that it offers full insulation protection. The fact that the steps of a ladder are above ground level does not mean that you are protected. Don't place yourself or an object you are using in a position where you or the object can contact a power line. Do not build a swimming pool beneath wiring of any kind, but especially not under power-lines.

Outdoor lights.   Be sure the lights you plan to use are approved for outdoors and are approved by a testing agency. The tag on an electric cord or an insignia on its box should have a safety endorsement. Never use indoor light bulbs outdoors.

Metallic trees.   Some artificial trees may have metallic leaves and branches. Never use electric lights on such trees. Use colored spotlights, but do not let the spotlights contact the tree.

Outdoor extension cords.   If you plan to use extension cords outdoors, make sure they are intended for that purpose *(NEC: Articles 240–4; page 109)*. The cord may or may not be equipped with a lamp at one end. Check the cord and the light for cracked insulation, bare wiring, or loose connections. The light should be enclosed in a cage that swings open to permit bulb changes. The bulb should be an outdoor type.

Grounding a television antenna.[5]   Television antennas are often installed by television technicians or hobbyists, and so NEC guidelines may not be observed. Also, such installations are often not under the jurisdiction of local electrical authorities.

---

[4]*(NEC: Article 100; page 29)*
[5]*(NEC: Article 810–15; page 866)*

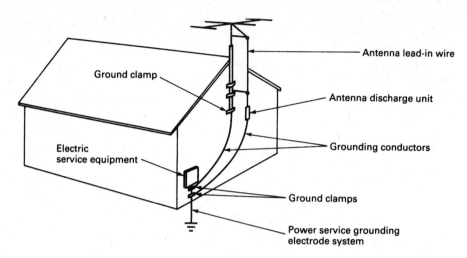

Figure 1–9.   Antenna grounding.

An outdoor TV antenna (Figure 1–9) system should follow certain precautions. (*NEC: Article 820–33; page 873*). These antennas should not be located in the vicinity of overhead power lines. When installing such a system, be careful to keep from contacting power lines or circuits since doing so is invariably fatal.

Be sure the antenna system is grounded so as to provide some protection against voltage surges and built-up static charges. *Section 810 (page 865)* of the NEC provides information with respect to proper grounding of the antenna mast and supporting structure, grounding of the lead-in wire to an antenna discharge unit, size of the grounding conductors, location of the antenna-discharge unit, connection to grounding electrodes, and the requirements of the grounding electrode.

Momentary Power Interruption

Momentary power interruptions (MPIs) are extremely brief electrical service interruptions usually lasting no longer than a few seconds. They result from temporary faults in electrical distribution. Basically, MPIs are caused by such factors as a lightning strike or a tree branch falling onto or brushing against a power line. One immediate consequence of an MPI is that digital electric clocks will start flashing, indicating that they need to be reset.

To minimize the nuisance of an MPI, select appliances with battery backup or carry-over systems when purchasing solid-state digital electronic products.

Users of personal computers should periodically store information in memory to prevent data loss.

How to Avoid Line Overloading[6]

One way of avoiding overloading a branch power line is to be aware of the effect of plugging in high-current devices. Power tools are often current-hungry devices, so it is always a good idea to read the manufacturer's data plate. Power tools for in-home use often have current ratings in the range of 2 to 13 amperes.

Usually, the larger and heavier the power tool, the greater the current demand. Before connecting such a tool, make sure no other current-demanding loads are plugged into the same line. If a tool requires a current of about 8 to 10 amperes, it can operate on the usual branch line, except for those that are connected to major appliances.

GROUNDING[7]

The words *ground* and *grounding* are widely used in electrical studies and work. They appear extensively in NEC articles beginning with the 250 series, page 120.

Function of Ground[8]

The ground system used in electrical work has several functions. (1) It is used as a reference point. All voltages, whether dc or ac, are measured between two points, one of which acts as a reference. The reference in electrical work is ground and is considered to be at zero potential with respect to a line that is 120 or 240 volts. Since the neutral wire is connected to ground, it is also a reference point for voltage measurements and is also at zero potential. (2) The use of a symbol to represent ground reduces the number of connecting lines in an electrical circuit diagram. (3) Grounding makes voltage measurements easier since a single ground point can be used as the reference for voltage measurements. If a voltmeter is used for testing, one test lead can end in a spring-loaded clip and can be fastened to a single ground point. The other test lead can then be used to probe a number of voltage points.

[6](*NEC Definitions: Article 100; page 24*)
[7](*NEC: Article 100; page 29*)
[8](*NEC: Article 100; page 30*)

Physical Ground

There are actually two physical grounds. One is located indoors, whether in a home, farm building, or factory, and requires a water supply, which flows, through a metal pipe that continues underground external to a building, thus providing the ground.

The other physical ground must be installed in the earth. The Code (*NEC: Article 250–53a; page 135*) refers to it as a *grounding electrode.* It consists of a metal pipe or rod with a $\frac{5}{8}$-inch diameter. Commonly, copper is used. A UL-listed rod is also available, carrying the tradename of CopperWeld, made of steel with copper fused to its surface. Although a lightning arrester is grounded, do not use this ground for the electric system. Another external ground could consist of a metal plate having a surface area of 2 square feet. Another possibility is the metal-sheathed underground portion of a building that is in direct contact with the earth or an underground tank. If more than one external ground is used, they must be bonded to each other to form a *grounding electrode system.* Assuming the non-availability of these grounding systems, a ground can be made by installing 10 feet or more of metal underground water pipe. The interior system ground must be bonded to the exterior ground system.

Bonding[9]

Bonding consists of joining conductive metals. Its function is to assure continuous electrical conductivity. Bonding can be used to join two or more metal parts of electrical equipment; it is referred to as an equipment bonding jumper. Such jumpers must be made of copper or other metallic substance that is corrosion resistant. The wire gauge of the bonding jumper is determined by the fuse or circuit-breaker current rating and must be equal to or greater than this rating (*NEC: Article 250; page 120*).

Ground and Neutral[10]

At one time the type of grounding used consisted of an existing metal water pipe system. This type of grounding was unsatisfactory since water pipes were often rusty, especially at their joints, resulting in poor conductivity.

The present interior grounding system is supplemented by an uninterrupted neutral wire continuing to the service box. Presently, the water pipe system is

[9]*(NEC: Article 250; page 120)*
[10]*(NEC: Article 236–22; page 91)*

being replaced wholly or partially by plastic tubing. The advantage of plastic tubing is that it is lighter than metal pipe, it is noncorrosive, and its sections are easier to join. However, it is unsuitable for use as a ground since it is nonconductive.

ELECTRICAL SAFETY

Electrical Safety in the Home

The fact that there have been no prior electrical problems is no guaranty that electrical safety will continue automatically. Adding electrical appliances decreases the margin of safety. Using multiple receptacles and taps or installing new appliances that have higher power requirements can push current requirements to their safety limits and beyond. It may be necessary to update branch power lines leading from the service box to the receptacle.

Warm receptacle or switch. A receptacle or switch that feels warm to the touch is evidence of an unsafe condition. The problem could be in the receptacle, switch, or wiring. The receptacle could also be overloaded, especially if it is the source of power for a multiple electrical strip.

Appliance line cords. Make sure that the line cords of plug-in appliances are in good condition. If the cord is frayed or worn, replace it. Covering the cord with electrician's tape is not recommended. If the associated switch needs to be jiggled to get it to work, replace it. If the appliance sparks, disconnect it immediately, replace it, or have it repaired. Do not use any appliance that supplies the characteristic sharp odor of ozone.

Polarized plugs. If an appliance plug is a polarized type, do not try to force it into a nonpolarized receptacle. Do not alter the prongs of the plug and do not replace it with a nonpolarized type. Replace the receptacle with one that is polarized.

Fire. In the event of an electrical appliance fire, remove its plug immediately. If you suspect a fire in a branch power line, disconnect it by opening its circuit breaker or removing its fuse.

Never use water on a live power line. Use a recommended fire extinguisher.

Most house fires are electrical. Most people are electrocuted by 120 volts, not 240 volts.

Water and electricity.   Water and electricity form a dangerous combination. Never turn on any power-line-operated appliance or tool if your hands are wet or if you are standing on a wet floor or damp ground. When working on a circuit breaker or fuse box, stand on a scrap rug, a rubber mat, or any other insulating material.

Do not poke or investigate the interior of an appliance when that appliance is connected to a receptacle. The fact that its switch is turned off is no guaranty of safety. Do not try to remove bread stuck in a toaster using a fork or your fingers without first disconnecting its plug.

Changing light bulbs.   When changing an electric light bulb, turn off the switch first and wait at least a minute to give the bulb a chance to cool. If the bulb has separated from its base, use work gloves to remove the base. A light bulb should turn easily into its base. If it does not do so, make sure you are not cross-threading the bulb base. Try a different bulb.

Extension and lamp cords.   Extension cords and lamp and appliance cords, although necessary, can be hazardous. Route them out of the way of vacuum cleaners and away from doorways, walking traffic, or chairs. They can present a tripping hazard. It is not advisable to route electrical cords under a rug. A better arrangement is to position a cord through a plastic channel mounted on a baseboard.

It helps to remember that lamp, appliance, and extension cords are part of the electrical wiring family, wiring that is normally concealed behind walls or enclosed in metal or plastic channels. Keep exposed wires away from children, pets, chemicals, oils, and heat.

Overloading.   The purpose of wiring is to carry an electrical current. If the wiring is overloaded, something that can be done by plugging in a current-hungry appliance, there may be a visible effect. A television picture may become smaller. Electric lights may get dim. These are both indications of overloading and call attention to the need for a branch wiring update, which means installing a heavier gauge of wire or running additional wiring in parallel with that of the existing branch.

Dedicated receptacles.[11]   Some appliances should have their own wiring branch going back to the fuse or circuit-breaker entry box. These include an electric stove, a washer, or a dryer. Since the electrical input to the entry box is 240 volts, that is the voltage that should be used. Make sure that is the recommended voltage indicated on the appliance.

[11]*(NEC: Article 100; page 33)*

Receptacles for power-operated tools. All receptacles for power tools should be three-blade types, and all plugs on these tools should be three-prong types. Do not use a three-prong to two-blade receptacle adapter for these tools. Always remove a connection from a receptacle by pulling on the plug. Do not pull the cord.

Electrical Safety and Power Tools

Not all tools operating from an ac receptacle are equally safe. A preferred safe tool is one having double insulation as a protection against electrical shock due to a fault in a branch electrical circuit. There are two ways of determining a tool's electrical safety. If the tool is doubly insulated, it will have a tag indicating that fact or else its operating instructions will emphasize this point.

Do not assume that a tool is electrically safe just because it uses a three-prong plug. A three-prong plug does supply a safe ground, but only if the branch circuit to which it is connected is also correctly grounded.

Double insulation is helpful, but it is no guarantee. In an environment with a high level of moisture, it does not prevent an electrical hazard. A better assurance is to use a GFCI (ground fault circuit interrupter). Again, this is no guarantee. Damp working conditions can make the tool unsafe.

When using an extension cord with a power tool, the ampacity of the cord must exceed the current requirements of the tool. To keep the plug of the power tool and the receptacle of the extension cord from separating, knot them. Loss of electrical power means stopping and restarting a tool with a job in progress and that can be hazardous.

Electrical Shock

A shock is the passage of an electrical current through the body. Simply making contact with a voltage source isn't enough to produce the sensation of a shock. For a shock to be physically evident the body must be in contact with the ground or with a voltage at a different voltage pressure level than the electrical part being touched. For a shock to exist the body must complete a circuit between two voltage sources at different electrical levels. Ground is commonly one of these points. The effects of the shock are more severe if the area of contact is large and stronger, in effect if the resistance at the contacts is lower.

Ground is a commonly used reference. Considering the wiring of a 120-volt receptacle, the electrical pressure between that receptacle and ground is 120-volts. Touching both of these satisfies the requirement for shock. Standing on an insulating pad when working with a voltage source is shock preventive.

There are various effects caused by a shock. In some cases the result is heat that is sufficient to burn body tissues, with areas subject to heat of as much as 2,000 degrees F. If the voltage source is AC the result can be a contraction of the muscles that can prevent pulling away from the voltage source.

Death can result from electrical shock within a few minutes, through interruption of the beating of the heart or the brain's control of breathing. These include paralysis, kidney damage, bleeding, and extensive tissue damage.

The maximum tolerable current is about 5 mA (5 milliamperes) but the heart can be affected by currents as low as 20 microamperes. A current as little as 10 to 20 mA can prevent the release of contact when the voltage is AC. A current of 30 mA can prevent breathing.

Capacitors connected across the filter section of a voltage rectifier circuit can hold a charge for a long time, even though the line voltage has been switched off. For this reason a high ohmage resistor, referred to as a bleeder, is used as a shunt to discharge such capacitors after the power supply has been turned off.

Sometimes an electrical shock can cause involuntary but rapid removal of the arm, resulting in some damage if an object is struck.

It is unwise to put electrically operated devices into a bathroom. If an appliance falls into a tub, death can be an invariable result whether the appliance is turned on or not.

Chapter 2

Basic Data
for Electricians

VOLTAGE

Voltage is probably the word most used by electricians for it is involved in every repair, installation, and component. But despite its extensive employment, it is often misunderstood or improperly used. You will hear phrases such as "this line carries 120 volts," with the implication that the 120 volts are gliding along a line, a conductor of electricity. That's just so much nonsense. Simply stated, *voltage* is electrical pressure.

Still another misconception is that the presence of voltage always means the existence of an electrical current. While voltage is essential for the establishment of a current, it is possible to have voltage without it. Voltage can exist across the terminals of an electric receptacle, but current does not flow until some device is plugged into that receptacle.

Another electrical myth is that a large voltage invariably means a large current. A car battery supplies a relatively low voltage but is often required to supply a substantial current, especially when starting the car. A night light in the home is connected across a much higher voltage than the headlights of a car, but requires far less current.

A voltage can be portable or fixed in position. A battery is portable; the voltage supplied by a receptacle is a fixed position source.

Various Names for Voltage

While the word *voltage* is commonly used, voltage is also known by a variety of other names, either spelled out completely or in the form of abbreviations. Voltage is also electrical potential, source potential, source voltage, potential difference (PD), electromotive force (emf), *IR* drop, line voltage, or electrical pressure. A 12-volt source can be said to have a potential of 12 or an emf of 12.

Voltage Representation

Letters can be used to represent voltage, particularly in formulas or on electrical components. These are V or v, E or e. Usually, following the formula, the selected letter (or letters) is identified by its function. An electric light bulb may be identified by the designation 120V.

Amount of Receptacle Voltage

The amount of voltage at a receptacle is controlled by a power utility. Generally, this electrical pressure remains constant, but during periods of heavy power usage, as in the summer when many air conditioners are functioning simultaneously, a utility may reduce the electrical pressure by a few percentage points.

Standardization of Line Voltage

The amount of ac power line voltage is approximately 120 and varies from one utility to the next. This variation is small enough so that different electrical appliances can be used nationwide. While standardization of line voltage would be desirable, certain factors make that difficult. The various electrical generators used by utilities do not all have the same voltage output, nor do they all have the same amount of regulation; that is, some have an output voltage that varies more with load than others. There is also a voltage drop along a power line, the amount depending on how close or how distant homes, offices, and factories are from the generators of the utility. Still another factor is the power output capability of the generator.

The American National Standards Institute (ANSI) has set up a standard for line voltage entitled American National Standard for Electric Power Systems and Equipment—Voltage Ratings (60 Hertz). This is covered by Standard ANSI C84.1 and specifies that receptacle voltage shall be 120 and 240 volts. The tolerance is 3 volts plus or minus and is for a three-wire system. Since receptacle voltage can vary depending on load, the ANSI standard is applicable to no-load voltage.

In homes and offices typical values of line voltage are 120 and 240 ac. The higher voltage is advantageous when current hungry appliances, such as washers and driers, are to be used. This is due to the fact that when using 240 volts current can be decreased by half.

Consider that the power requirement of an appliance is based on:

$$P = E \times I$$

Here P is the power in watts, E is the voltage in volts, and I is the current amperes. If the voltage is doubled, as it will be when it is increased from 120 to 240 ac, the same amount of power will be supplied if the current is reduced by half. The advantage is a permissible reduction in the wire gauge being used to supply current.

UNITS OF VOLTAGE

The basic unit of voltage is the *volt*. The range of voltage is enormous. It can be as little as a few millionths of a volt to millions of volts. That range is over a trillion units, or the number 1 followed by 12 zeros. The basic unit is 1 volt. All other voltages, smaller or larger than 1 volt, use the volt as a reference.

Microvolt

One microvolt is one-millionth of a volt. Arithmetically, a microvolt is 1/1,000,000 volt or 1 volt divided by 1 million. It takes 1 million microvolts to equal 1 volt.

$$1 \text{ volt} = 1 \times 1,000,000 = 1,000,000 \text{ microvolts}$$

The abbreviation μV is used for microvolts. μ is an abbreviation for micro; V represents volts. Thus, 1 V = 1,000,000 μV.

It is sometimes necessary to convert microvolts to volts. Multiplication is used in the change from volts to microvolts. Division is used for the reverse process, that is, converting microvolts to volts.

Microvolt to volt conversion. To convert microvolts to volts, divide microvolts by 1,000,000.

$$1 \text{ microvolt} = \frac{1}{1,000,000} \text{ volt} = 0.000001 \text{ volt}$$

or

$$1 \text{ μV} = \frac{1}{1,000,000} \text{ V} = 0.000001 \text{ V}$$

Millivolt

One millivolt is one-thousandth of a volt. The millivolt can be abbreviated as mV, where m = milli and V = volts. The relationship of millivolts and volts can be written as follows:

$$1 \text{ millivolt} = \frac{1}{1000} \text{ volt} = 0.001 \text{ volt}$$

$$1 \text{ mV} = \frac{1}{1000} \text{ V} = 0.001 \text{ V}$$

$$1 \text{ volt} = 1000 \text{ millivolts}$$

$$1 \text{ V} = 1000 \text{ mV}$$

Relationships of Millivolts and Microvolts

One millivolt is equal to 1000 microvolts.

$$1 \text{ mV} = 1000 \text{ }\mu\text{V}$$

$$1 \text{ }\mu\text{V} = \frac{1}{1000} \text{ mV} = 0.001 \text{ mV}$$

One kilovolt, abbreviated as kV, is one-thousand volts. Its relationship to volts, millivolts, and microvolts can be stated as follows:

$$1 \text{ kilovolt} = 1000 \text{ volts} = 1/1000 \text{ megavolt} = 0.001 \text{ megavolt}$$
$$= 1,000,000 \text{ millivolts} = 1,000,000,000 \text{ microvolts}$$

It is sometimes difficult to work with voltage conversions since large decimal numbers can be involved. Electrical problems often supply data in terms of basic units, not only for voltage, but for current and resistance as well. Sometimes the numbers supplied are in terms of multiples or submultiples, and before any solution can be attempted, it is necessary to convert multiples and submultiples to basic units.

CURRENT

The electrical pressure of a voltage source can produce the movement of extremely small particles known as electrons. The sum of this movement is called an *electrical current,* but it is more often simply referred to as a *current.*

There are two basic types of current, dc and ac. DC is an abbreviation for direct current; ac stands for alternating current. A direct current can be supplied by batteries, an electronic power supply, or an electromechanical device called a dc generator, or a dynamo. AC, such as that supplied by an electric utility, is produced by an ac generator.

The basic difference between dc and ac is in their direction of current flow. DC, as its name implies, always has its current flow in one direction and is represented as in the diagram in Figure 2–1. The circuit consists of three components: a battery or source voltage, a single-pole, single-throw on–off switch, and a small electric light bulb, with these parts connected by wire. The arrows indicate the direction of current flow. Just as no one part of a wheel moves before or later than any other part, all the current moves simultaneously. However, it is a convenient fiction to consider the current as starting at the negative (minus) terminal of the battery. The direction of current is through the closed switch and then the bulb. The current continues to the plus terminal of the battery and then through the interior of the battery, emerging once again from the negative terminal. The direction of flow of the current can be changed by transposing the leads to the terminals of the battery.

Two-wire Line

The circuit in Figure 2–1 uses two wires. The amount of current flowing through each wire is always identical. There is no accumulation of current, and the amount of current leaving the negative terminal of the battery is always the same as that returning to its plus terminal. It is also equal to the amount of current flowing through the battery.

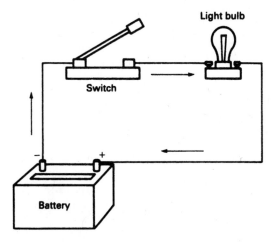

Figure 2–1. Arrows indicate direction of current flow when switch is closed.

UNITS OF CURRENT

The basic unit of current is the *ampere,* which is the reference for all other current units. The ampere is represented by the letter A: thus 6 A = 6 amperes.

Milliampere

One milliampere is equivalent to one-thousandth of an ampere. Thus

$$1 \text{ milliampere} = \frac{1}{1000} \text{ampere} = 0.001 \text{ ampere}$$

$$1 \text{ mA} = \frac{1}{1000} \text{A} = 0.001 \text{ A}$$

$$1 \text{ ampere} = 1000 \text{ milliamperes}$$

$$1 \text{ A} = 1000 \text{ mA}$$

Reading a Direct-current Graph

A graph is a pictorial representation of what is happening in a circuit, such as the drawing in Figure 2–2. The horizontal base line represents time, which in this example is measured in seconds. The vertical line at the left is the amount of current flow in amperes.

According to this graph, the current flow reaches its maximum practically im-

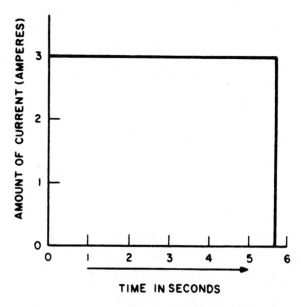

Figure 2–2. Graph of a direct current.

mediately. The amount of current flow is 3 amperes and exists for $5\frac{3}{4}$ seconds. At the end of that time the circuit switch is opened and current flow becomes zero.

ALTERNATING VOLTAGE

There are two basic differences between an alternating voltage, abbreviated as ac, and a dc voltage. Unlike a dc voltage, ac reverses its polarity. And, again unlike a dc voltage, ac is not a fixed amount but changes with time.

Figure 2–3 is a graph of an alternating voltage. It shows a line that starts at zero and, unlike the graph of a dc voltage, rises gradually to a peak and then decreases just as gradually to zero. At zero the line continues, but downward this time, until it reaches a peak, gradually, and finally returns to zero. This behavior of an alternating voltage keeps repeating again and again. The graph consists of two parts, a section that moves upward and a downward section. The upward-moving section represents the voltage when its polarity is positive; the downward section represents the polarity when it is negative. In other words, an ac voltage is one whose polarity keeps changing. We could change the polarity of a dc voltage source by transposing the connections mechanically.

Since the direction of current flow depends on the polarity of the voltage source, an alternating voltage generates a current that flows first in one direction and then reverses itself, forcing the current to flow in the opposite direction.

This is just one of the effects of an alternating voltage. The voltage does not remain constant but keeps changing all during the time the voltage is being generated. This is the type of voltage that is produced by electric utilities and that is available at customers' receptacles.

So we have two types of voltage, ac and dc. We cannot say one is more important than the other or better. They are different, both are necessary, both are widely used, and electricians must know both and be able to control both.

The horizontal line that divides the upper half from the lower half in the

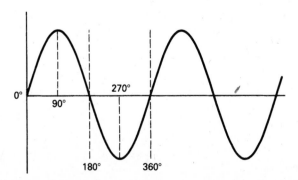

Figure 2–3. Graph of an ac voltage.

graph of Figure 2–3 can be marked in time units in the same way as the dc graph, but commonly it is indicated in degrees. The starting point is zero degrees (0°), the center point is 180°, and the end point is 360°. Note that the voltage is zero at three different times: 0°, 180°, and 360°. This graph of the behavior of an ac voltage is called a *wave* simply because it looks like one. The voltage that is delivered by an electric utility is zero three times when considering just a single wave.

Measuring AC Voltage

Measuring a dc voltage is simple since that voltage remains fairly constant for a new or freshly charged battery. An ac voltage, though, varies from zero to a peak and from that peak back to zero again, and it does so continuously. This presents the problem of how to measure something that keeps changing. As a result, a number of ways are used to determine the amount of an ac voltage.

   Peak AC voltage.   An obvious way of measuring an ac voltage is to determine its maximum or peak value (Figure 2–4). This might seem a good solution to the problem except that the ac waveform has only two peaks, one at 90° and the other at 270°. We can control the amount of peak voltage since we can adjust the source, an ac generator, to supply the amount of voltage output. As a start, assume we make that voltage equal to 1. We can write this as

$$\text{peak voltage} = 1$$

or

$$\text{ac peak} = 1$$

   Peak-to-peak.   If the peak voltage is 1, then the electrical pressure between the positive peak to the negative peak (Figure 2–5) is twice this amount, that is, 2.

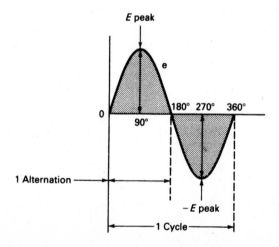

Figure 2–4. This ac waveform has a positive and a negative peak.

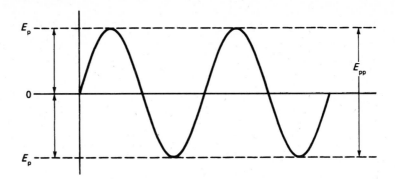

Figure 2–5. The peak-to-peak voltage is measured between the positive peak and the negative peak.

$$\text{peak-to-peak voltage} = 2$$
$$\text{emf p–p} = 2$$

If, as an example, the peak voltage should happen to be 185 volts, then the peak-to-peak voltage would be 2×185 or 370 volts.

Instantaneous values of ac voltage. This is the value of an ac voltage at any moment in time. The peak voltage can be considered as instantaneous since it is measured only at the 90° and 270° points. As instantaneous voltage (Figure 2–6) can be measured at any degree point along the base line, but that degree point should be mentioned when discussing any instantaneous value.

Average value of an ac wave. The average value of the wave shown in Figure 2–7 is a summation of all the instantaneous values. It is zero since there are an equal number of positive and negative instantaneous values. Consequently, the average value is considered only over one half-cycle of the wave, that is, from its start or 0° to 180°. The average value can be obtained by dividing 2 by pi (π), with the value of pi taken to be 3.14159265. When this arithmetic is done, the average value is found to be 0.636619, but this is generally rounded off to 0.637. If the peak value of an ac wave is 110 volts, then the average value is 70.07 volts. This is obtained by multiplying the peak value by 0.637.

$$\text{peak value} \times 0.637 = \text{average value}$$

and, in this example,

$$110 \times 0.637 = 70.07$$

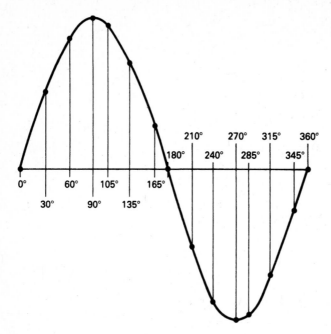

Figure 2–6. Instantaneous values of an ac wave.

RMS value of an ac wave.   The rms, root-mean square value (also known as the *effective value*) is probably the most widely used measurement for the type of ac wave shown in Figure 2–3 and is a type of instantaneous value averaging. The rms value is obtained by dividing the digit 1 by the square root of 2 and is shown as

$$\text{rms value} = \frac{1}{\sqrt{2}}$$

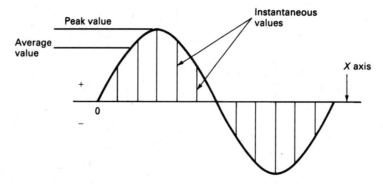

Figure 2–7. The average value is 63.7% of either the positive peak or the negative peak.

The square root of 2 is equal to 1.41423 and can be written as

$$\sqrt{2} = 1.41423$$

When this result is divided into 1, we get

$$\frac{1}{1.41423} = 0.707107$$

but this answer is usually rounded off to 0.707. Multiplying this number by the peak value supplies the rms value (Figure 2–8).

The basic line voltage commonly used in the home or office is about 120 volts. When this number is divided by 0.707, the result is the peak voltage. Thus

$$\frac{120}{1.707} = 169.73 \text{ V peak}$$

This peak voltage arrives twice during the completion of each ac wave. When power is turned on to an electric device such as a light bulb, the effective voltage reaching that bulb depends on the instantaneous voltage. If it happens to be 169 volts at that moment, it is possible for the bulb to flash and burn out. If the bulb continues to function, it operates over the effective value of the applied voltage, which is approximately 120 volts.

Table 2–1 supplies data on calculating average, effective, peak, and peak-to-peak values. The multiplication factor is supplied for all these.

Types of Waves

There are many types of ac waves, which are named for their shape. An ac wave can be square, rectangular, or triangular, or it may have no particular shape at all. The wave shown in Figure 2–3 is referred to as a sine wave and is the type of

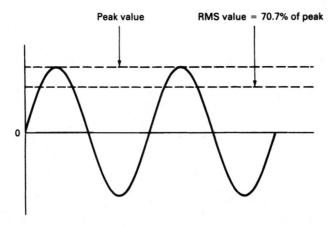

Figure 2–8. The rms value is 70.7% of the peak. This is the amount of voltage supplied to receptacles.

TABLE 2–1. CALCULATION OF SINE-WAVE VALUES

| | Multiply by this value to get: | | | |
Given this value:	Average	Effective	Peak	P-P
Average	—	1.11	1.57	1.274
Effective	0.9	—	1.414	2.828
Peak	0.637	0.707	—	2.0
P-P	0.3185	0.3535	0.50	—

voltage wave supplied by electric power companies. The various ways of measuring an ac wave that have been described apply to sine waves only.

Sine Waves of Voltage and Current

A voltage sine wave produces an electrical pressure that starts at zero, has a positive and a negative peak, and ends at zero. When sent through an electrical device such as a light bulb, this voltage results in a current having the same shape as the voltage producing it; it is known as a current sine wave. All the measurements used for a sine wave of voltage can be applied to this current.

RESISTANCE

All currents, whether dc or ac, encounter resistance when flowing through a conductive element. This element can be a wire, a soldered connection, a mechanical connection, a metal plate, the human body, and so on. Ordinarily, it is desirable to have the resistance as low as possible.

Units of Resistance

The basic unit of resistance is the *ohm*. Fractions of an ohm are usually not a consideration in electrical work. Multiples, such as the kilohm (one thousand ohms) and the megohm (one million ohms), are available. Thus, 5 kilohms is 5×1000 ohms. Kilohms divided by 1000 is equal to ohms; 6.4 kilohms divided by $1000 = 6.4$ ohms.

The megohm is equivalent to ohms divided by 1 million. 6,500,000 ohms divided by 1,000,000 equals 6.5 megohms. To convert megohms to ohms, multiply megohms by 1,000,000. Thus, 4.3 megohms equals $4.3 \times 1,000,000$ equals 4,300,000 ohms.

Letter Symbols

Letter symbols are frequently used in formulas as an easy way of representing electrical quantities. E is used for voltage; I for current, and R for resistance. When describing a specific resistance, the Greek letter omega (Ω) can be used. As an example, a wire has a resistance of 1.7 ohms, or 1.7 Ω. For values greater than 1000 ohms, the letter k (for kilo-) is used as a multiplier. As an example, the primary winding of a transformer can be indicated as having a resistance of 1100 ohms, or 1.1 kilohms or 1.1 kΩ.

The letter abbreviation of mega is M. An extremely narrow, very long conductor may have a resistance of 1.1 MΩ.

OHM'S LAW

There is a relationship between voltage, current, and resistance for circuits involving dc. Known as Ohm's law, it can be written as

$$\text{voltage} = \text{current} \times \text{resistance}$$

The formula can be simplified by substituting letters for words, thus making the formula easier to remember and easier to manipulate. In terms of letters,

$$E = I \times R$$

The formula assumes that all quantities are in basic units: E is in volts, I is in amperes, and R is in ohms. If any one or more of these are in multiples or submultiples, they must first be converted to basic units. Thus, if I is in amperes and R is in kilohms, R must first be converted to its value in ohms.

The Ohm's law formula can be used to find resistance or current as well. Thus,

$$E = I \times R$$

$$R = \frac{E}{I}$$

$$I = \frac{E}{R}$$

A simple device for memorizing Ohm's law appears in Figure 2–9. Cover the unknown quantity with a finger. To determine the value of E, look below that letter and the result is I next to R, that is, I times R. To find the value of I, cover it and what remains is E over R, or E divided by R. Similarly, to learn the value of R, put a finger over that letter and what remains is E over I, or E divided by I.

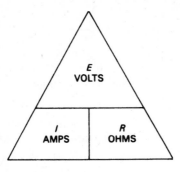

Figure 2–9. Memory aid for Ohm's law.

POWER

Three factors govern the cost of using any electrical appliance: the amount of voltage it needs, its current requirements, and the length of time the appliance is used. The product of two of these, current and voltage, results in a new unit known as the *watt* (Figure 2–10).

To determine the wattage of an appliance such as a toaster or an electric light bulb, multiply its current in amperes by the voltage of the line. A toaster that uses 2 amperes when connected to a 120-volt line has a power rating of $2 \times 120 = 240$ watts. The wattage rating is often marked on the data plate of the appliance.

The amount of power in watts can be calculated from

$$W = E \times I$$

which is sometimes written as

$$P = E \times I$$

As in the case of Ohm's law, this formula can supply us with two other formulas:

$$E = \frac{W}{I}$$

and

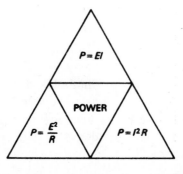

Figure 2–10. Formulas for calculating power in resistive ac circuits.

$$I = \frac{W}{E}$$

For solving these formulas, use basic units only.

Various other formulas can be used to calculate the power used by an electrical device. One of these is

$$W = I \times I \times R$$

Since $I \times I = I^2$, the formula can be written as

$$W = I^2 R$$

Watt

The basic unit of electrical power is the *watt* (W). The watt has submultiples (power less than 1 watt) and multiples (power greater than 1 watt).

Kilowatt

A kilowatt (kW) is a thousand watts. To convert watts to kilowatts, divide watts by 1000. 650 watts divided by $1000 = 6.5$ kilowatts, or 6.5 kW.

Relationship between Current and Voltage

A high-voltage source does not necessarily mean a large amount of current. As a general rule, a high voltage often means a low current, while a low voltage is sometimes associated with large currents. But there are many exceptions. Consider the statement that high voltage means a low current not as a rule, but rather as a possibility.

Amount of Loading

An electric utility generator has a very high current capability and is able to deliver as much current as required by its customers. At night, with all appliances and lights turned off, the current needs of a single home are quite small and might total less than 1 ampere. The load could possibly be that required by one or more electric clocks and a few night lights. With the start of the day, various appliances may be used: a toaster, room lights, an electric heater, a broiler, an air conditioner, a heating unit, and so on. While the voltage at all the receptacles remains fairly constant, the current demand has increased. This current demand represents the *load* and is variable.

Total Power

Since the voltage at receptacles is constant, the total power being used at any time can be calculated by adding all the currents being supplied to the various appliances and then multiplying this sum by 120, the approximate power-line voltage. If a toaster takes 6 amperes, a color TV 3 amperes, and an air conditioner 8 amperes, then the total current demand is 6 + 3 + 8 = 17 amperes. And 17 × 120 = 2040 watts, just a little more than 2 kilowatts.

Appliance Power

To learn how much power is required by each appliance, look at the data plate on each unit. If there is no data plate, consult the instruction manual that accompanied the appliance or write to the manufacturer. The information is available. It is just a question of digging it out. In some instances the appliance may have a current, but no power rating, on its data plate. Multiply the figure supplied by the receptacle voltage to get an approximation of the wattage of the unit.

Two voltages are available in a residence: 120 and 240 volts. Appliances using 240 volts are current-hungry types. Those operating from 120-volts demand much less current.

The power used by an appliance is the product of voltage and current, that is, the voltage (in volts) multiplied by the amount of current (in amperes). In terms of a formula,

$$P = E \times I$$

If an appliance operates from a 120-volt line and requires 2 amperes of current, its power demand will be $P = E \times I = 120 \times 2 = 240$ watts. It would be possible to redesign the appliance to work from half the original voltage, that is, $\frac{120}{2} = 60$ volts. But since the appliance requires 240 watts of power, the amount of current would need to be doubled.

This presents a problem. Increasing the current flow means a heavier-gauge wire will be needed to carry the current. Not only is this more expensive, but the wire is more difficult to handle. Note, though, that the cost of operating the appliance remains the same and is 240 watts in both instances.

It is possible to move in the other direction and to increase the voltage to 240. In that case the current requirement would be reduced to 1 ampere. Once again, $P = E \times I = 240 \times 1 = 240$ watts. With this reduced amount of current, a thinner-gauge connecting wire can be used. This reduces the cost of the wire and makes labor easier as well. This is why a number of appliances, such as clothes washing machines, clothes dryers, and electric stoves, are 240-volt types.

Cost of Power

An electric utility not only counts the total wattage but also the amount of time of such usage. The basic unit for measuring cost is the kilowatt hour. There is no standard for such charges, and they can vary from one geographical area to another. Electrical charges are sometimes based on a sliding scale: the more power that is used, the lower the cost per kilowatt hour.

While costs per kilowatt hour may be measured in pennies, a monthly electric bill can sometimes be staggering. There are two reasons for this. The first is that there are 24 hours a day and at all times, during the day or night, electrical power is being used.

Multiply 24 by 31 days in a month to get the maximum possible time usage and the result is 744 hours. If a home has twenty 60-watt electric light bulbs, the total power consumption if all are turned on at the same time would be $60 \times 20 = 1200$ watts or 1.2 kilowatts. This would be a maximum figure since the lights would not all be on at the same time nor would they usually be on at night. At the same time, though, some of the bulbs might be rated at 75 or 100 watts, and the total power might include heat lamps, and outdoor flood lights. In addition, appliances such as an electric toaster, broiler, a plug-in heating unit, and a microwave oven, might be turned on, and so there is a good possibility for high-wattage usage. The criterion for electrical power cost depends on two factors: the amount of current electrical appliances require and the length of time that they remain turned on.

Table 2–2 lists the average wattage of various appliances. This is not a complete list and the wattage ratings are approximations. The table permits a comparison between various appliances and supplies an indication of which appliances help boost an electric bill. A quick look at the data plate on an appliance should supply a rough idea of what its usage cost will be. Often, the operating cost of an appliance, over a period of time, will be larger than the purchase price.

KILOWATT-HOUR METER

Most people do not check their electric bills, accepting them on faith, correctly reasoning that power companies are not in business to cheat them. But an electric power bill is based on a reading of a kilowatt-hour meter somewhere inside or outside the home. The meter reader is a professional and is experienced, but is also human and can make mistakes.

The meter that measures the use of electric power is called a kilowatt-hour meter. It makes two measurements: the power used in watts and the elapsed time

TABLE 2–2.   WATTAGE RATINGS FOR VARIOUS APPLIANCES

Appliance	Average wattage for a 120-volt line	Appliance	Average wattage for a 120-volt line
Food Preparation		**Comfort Conditioning** (*cont'd*)	
Barbecue Grill	1500	Dehumidifier	257
Blender	300	Electric Blanket	177
Broiler	1140	Fan (attic)	370
Coffee Maker	1200	Fan (circulating)	88
Deep Fryer	1448	Fan (rollaway)	171
Dishwasher	1200	Fan (window)	200
Food Processor	480	Furnace Fan	500
Frying Pan	1196	Heater (portable)	1322
Hot Plate	1200	Heating Pad	65
Mixer	27	Humidifier	177
Oven, Microwave	1450	**Health and Beauty**	
Range, with Oven	12,200	Hair Dryer	600
Sandwich Grille	1161	Heat Lamp (infrared)	250
Toaster	1146	Shaver	15
Trash Compactor	400	Sun Lamp	279
Waffle Iron	1200	Toothbrush	1
Waste Disposal	445	**Home Entertainment**	
Food Preservation		Radio	71
Manual Defrost (16 cu. ft.)	63	Television (color, solid state)	145
Automatic Defrost	93	VCR	25
Refrigerator/Freezer		**Housewares**	
Manual Defrost (12.5 cu. ft.)	66	Indoor	
Automatic Defrost	93	Clock	2
Laundry		Floor Polisher	305
Dryer	4856	Sewing Machine	75
Iron (hand)	1100	Vacuum Cleaner	630
Washing Machine	512	Outdoor	
Water Heater	2475	Lawn Mower	75
Comfort Conditioning		Leaf Blower	72
Air Cleaner	50	Weed Trimmer	72
Air Conditioner (room)	860		

(Courtesy Jersey Central Power and Light Company)

of usage. There are various kinds. One is a cyclometer type and looks like a small box with rectangular openings through which numbers can be seen. The cyclometer supplies a direct reading in kilowatt-hours and is as easy to read as a digital clock. When the meter reaches 9999 it begins again with 0.000. Figure 2–11 is a drawing of such a meter.

      Readings of the meter can be made in several ways. Select a particular hour of a certain day and take a reading. Then, one month later, at the same hour and

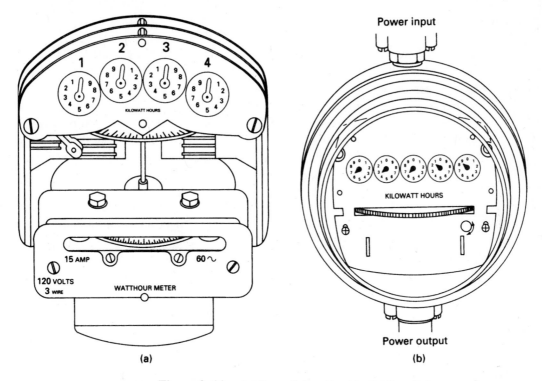

Figure 2–11.   (a) Four-dial wattmeter; (b) five dial.

on the same day, take the next reading. Subtract the first reading from the second, supplying the power usage for that month. Two facts will emerge if this is done month after month. During certain months of the year, power usage may rise dramatically. The second is that it supplies a yardstick, for comparing the results with those on the electric bill.

To make a more accurate check, take a reading immediately after that taken by the meter reader working for the power utility. If, for example, on March 1 the kilowatt-hour meter reads 1657 and a month later it reads 5003, the power consumption is

$$5003 - 1657 = 3346 \text{ kilowatt-hours}$$

Multiply this number by the prevailing rate per kilowatt-hour, add taxes, and the result should be close to that on the electric bill.

The cyclometer dial is simple and supplies the power usage directly. The first dial measures thousands, the next dial measures hundreds, the third dial measures tens, and the last dial, the one at the right, measures units or ones. Some cyclometers have four dials; others have five.

FREQUENCY

Although there are a number of different types of waveforms, the one used by electric utilities is the sine wave. This wave consists of two halves that are mirror images of each other. The upper half extends from 0° to 180° and is commonly taken to represent a positive voltage. The lower half, from 180° to 360°, is a negative voltage. Each half is called an *alternation*. Taken together, the two halves form a single complete cycle. The number of cycles per second is the frequency of the wave and is expressed in *hertz* (Hz). The usual frequency of the ac power line is 60 hertz. At one time, frequencies of 25 and 50 Hz were used, but 60 Hz is now the standard. With this frequency a single cycle is completed in $\frac{1}{60}$ second. Frequency is applicable to ac only, and current in a dc circuit is considered as 0 Hz.

Frequency is part of the data supplied on the data plates of appliances. An appliance marked 200 W, 120 V, 60 Hz indicates that the device must be connected to a receptacle supplied with 120 V ac whose frequency is 60 Hz. Under these conditions the appliance, when operative, will utilize 200 watts of electrical power.

An appliance such as an electric clock is frequency dependent. It will supply accurate time only as long as the line frequency is 60 Hz or averages this frequency over a period of time, such as 24 hours. Most utilities supply line voltage having a remarkably accurate frequency. Other appliances, such as toasters or broilers, are not frequency dependent. Radio and TV sets require a 60-Hz input.

Frequency Multiples

The hertz is the basic unit of frequency, and multiples should be converted to basic units before being used in a formula. Multiples of the hertz are commonly used in electronics but rarely in electrical work. The kilohertz is equivalent to 1000 Hz and the megahertz is 1,000,000 Hz.

PHASE

The word *phase* is commonly used when working with ac circuits, with transformers, and with motors. Phase is related to time and can be specified either in units of time or degrees.

Phase may sound complicated, but all it means is the starting time relationship of a voltage and a current, or a pair of currents, or a pair (or more) of voltages. While the concept of phase is common in ac circuits, it is much less so in dc.

When a dc circuit is turned on, both the voltage and the current operate simultaneously. The voltage increases to a peak practically instantaneously, and the current increases at the same time. When shown on a graph, the beginning of the voltage and of the current are the same and are at 0° on the graph. Since there is no time separation between the current and voltage, the two are in phase. In terms of degrees, the phase is 0.

With an ac circuit that consists of an ac voltage source and a coil, the current will lag the voltage by a maximum of 90°. This is referred to as the *phase angle.* The actual amount of the phase angle is usually less than 90°, depending on the structure of the coil winding.

TIME

To the important factors in electric circuits of voltage, current, and resistance, we can add time. Unlike dc circuits, in which voltage and current are in step, that is, start and stop simultaneously, in ac circuits, current and voltage are not necessarily in step. The time separation of voltage and current is covered by the single word *phase.* Voltages and currents that begin and finish with no time separation are in phase, and those that are out of step are out of phase.

Not only can an ac voltage be out of phase with its accompanying current, but a pair of currents or a pair of voltages can be in phase or out of phase.

Figure 2–12 shows a voltage and a resulting current that are out of phase. Since, in terms of time, the voltage starts before the current, the relationship can be described in either of two ways. We can say that the current *lags* the voltage or the voltage *leads* the current, with both statements having an equivalent meaning. The *base line,* the line which divides the positive half of the wave from the negative half, can be marked in degrees, with this measurement starting at 0° and ending at 360°, or it can be marked in time units, using seconds.

It is also possible for the current to precede the voltage, in which case the

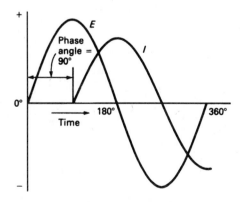

Figure 2–12. In a purely inductive circuit the current lags the voltage by 90°.

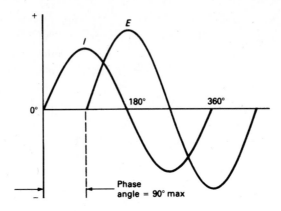

Figure 2–13. In a purely capacitive circuit the current leads the voltage by 90°.

behavior is described as one in which the current leads the voltage (Figure 2–13) or in which the voltage lags the current.

In those cases in which the current and the voltage both start and stop at the same time, the two, the current and the voltage, are in step or in phase (Figure 2–14).

These voltage and current relationships have nothing to do with current or voltage amplitude. The current can be stronger than the voltage, or the voltage can be stronger than the current.

Time is used in electric circuits as a unit of measurement. A relay may be designed to open or close after a specific amount of time; the rotation of the armature of a motor or generator may be measured in time units; a fuse may be intended to open after a certain amount of time following an overload. A sine wave may be measured by determining the number of complete cycles per unit of time.

Time can be calculated in terms of hours, minutes, or seconds. One hour is

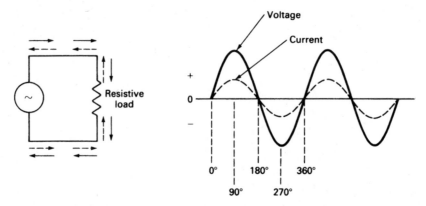

Figure 2–14. Voltage and current are in phase in a purely resistive circuit.

equal to 60 minutes, and one minute is equal to 60 seconds. The abbreviation for hours is h, and for seconds it is s. Time units in electric circuits are either in seconds or hours, but rarely in minutes.

$$1 \text{ h} = 60 \text{ minutes} = 3600 \text{ s}$$

$$1 \text{ minute} = \frac{1}{60} h = 60 \text{ s}$$

$$1 \text{ s} = \frac{1}{60} \text{ minute} = \frac{1}{3600} \text{ h}$$

The abbreviation for time is the letter t. Thus, time in minutes equals 60 seconds can be written as:

$$t = 60 \text{ s}$$

INDUCTANCE

Every current, whether ac or dc, is accompanied by a magnetic field. If the current is dc, the magnetic field remains steady except when the current starts and when it stops. If the current flows through a wire, the wire behaves as though it were a magnet. When the current is switched off, the wire loses its magnetism.

When an alternating current flows through a wire, the wire also behaves like a magnet. Several important effects take place with this kind of current. Since the current keeps changing in strength, the magnetism surrounding the conductor is in step with the current flowing through it. It starts at zero and reaches its peak strength when the current is strongest. When the current reverses its direction, the magnetism changes it polarity.

When a magnetic field surrounds a wire and that magnetic field keeps changing its strength, it induces a voltage across the wire, with this voltage in opposition to the original applied voltage. We now have two voltages across the conductor: the original ac voltage and an induced voltage. Since these two voltages oppose each other, the effect is to reduce the amount of the original current, just as though we had used a resistor in the circuit. This additional invisible resistance is called *reactance* and is represented by the letter X. The reactance of a coil is measured in ohms, and its symbols are the same as for ohms. Reactance can be in kilohms as well as megohms.

The ability of a conductor to have a magnetic field when a current flows through it is called its *inductance* and is represented by the letter L. If the inductance is wound in the form of a coil instead of being straight, its inductance increases. Inductance is variable and depends on the physical construction of the coil. The greater the number of coil turns is, the greater the inductance. Inductance also increases if the coil is wound on a form and some type of iron is inserted into that form. Inductance also depends on how close (or separated) the coil turns are from each other, the diameter of the coil, on the number of layers of turns of wire, and the shape of the coil. These factors supply considerable control over the amount of inductance of the coil. That, in turn, helps determine the reactance and consequently the amount of current permitted to flow through the coil.

Units of Inductance

The basic unit of inductance is the *henry* (H). Since the henry is a large unit, there are no multiples. Submultiples, though, are common and include the millihenry (mH) and the microhenry (μH). Their relationships can be expressed as

$$1 \text{ mH} = 0.001 \text{ H} = \frac{1}{1000} \text{ H}$$

$$1 \text{ H} = 1000 \text{ mH}$$

The relationships of the microhenry and the henry are

$$1 \text{ μH} = 0.000001 \text{ H} = \frac{1}{1,000,000} \text{ H}$$

$$1 \text{ H} = 1,000,000 \text{ μH}$$

It is sometimes necessary to work between millihenrys and microhenrys. These can be expressed as

$$1 \text{ mH} = 1000 \text{ μH}$$

$$1 \text{ μH} = 0.001 \text{ mH} = \frac{1}{1000} \text{ mH}$$

The resistance of a coil is sometimes referred to as its dc resistance. It is the resistance the coil would have when measured with a test instrument such as an ohmmeter or the resistance it would have in a dc circuit.

INDUCTIVE REACTANCE

Since a coil is also known as an inductor, its reactance or opposition to the flow of an alternating current is known as *inductive reactance*. It is identified in this way since there are other components that also have reactance. The symbol for inductive reactance is X_L. X is the symbol for reactance in general, and the subscript L identifies the type of reactance, which in this case is inductive.

   The amount of inductive reactance depends on a number of factors, including the frequency of the ac and the inductance of the coil. In terms of a formula, it is written as

$$X_L = 2\pi f L$$

   In this formula 2π is equal to 6.28, f is the frequency of the ac in hertz, and L is the inductance in henrys. If the amount of inductance is supplied in millihenrys or microhenrys, these must first be converted to the basic unit, or henrys. The frequency is usually (but not always) equal to 60 Hz. For the most part, then, the reactance of a coil is influenced by its inductance. This is controlled, as indicated earlier, by the physical characteristics of the coil.

AC POWER

As indicated earlier, power in a dc circuit is simply a matter of current, voltage, and resistance. But in an ac circuit, the fact that the current and voltage can be out of phase reduces the amount of available power delivered to a load; this condition is described by a term known as *power factor* (pf).

Impedance

Often, in an ac circuit, a coil and a resistor will be connected as indicated in Figure 2–15, known as a *series circuit*. The resistance supplies some opposition to the flow of current and so does the inductive reactance, with both units, the resistance and reactance, measured in ohms.

   The resistance and reactance cannot be added directly. The total opposition to the movement of current is called *impedance,* and this total opposition can be measured by the formula

$$Z = \sqrt{R^2 + (X_L)^2}$$

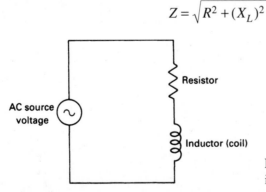

Figure 2–15.   Series resistor and inductor circuit.

This formula tells us to add the resistance squared and the inductive reactance squared and then to take the square root of this sum.

CAPACITORS[1]

Basically, the components used in electrical work are simple. A coil, for example, is just a length of wire wound in a circular manner. A *capacitor* is another component and in its elementary form consists of a pair of metal plates facing, but not touching, each other. Its distinctive feature is that it is capable of holding an electrical charge. The material between the plates can be air or some substance such as mica, paper, or plastic film. Whatever it may be, it is referred to as a *dielectric*. Figure 2–16 shows the fundamental arrangement of a resistor-capacitor series circuit.

Capacitance

The ability of a capacitor to store a charge is measured in a basic unit referred to as the *farad*. This is an extremely large unit and so submultiples are commonly used. A more practical unit is the microfarad (μF), which is equal in value to one-millionth of a farad. But sometimes even this submultiple is too large, and so a still smaller unit, the picofarad (pF) is used. A picofarad is one-millionth of a microfarad

$$1 \text{ F} = 1,000,000 \text{ } \mu\text{F} = 1,000,000,000,000 \text{ pF}$$

The amount of capacitance is determined by the area of the plates that form the capacitor, by the type of dielectric, and by the distance the plates are from each other. When put to work, a capacitor is placed across a voltage from which it obtains an electrical charge. The ability of a capacitor to withstand the voltage put across its terminals is known as its *working voltage* (WV).

WV depends on the way the capacitor is manufactured, the type and thickness of the dielectric, and the temperature. WV is often marked on the outside

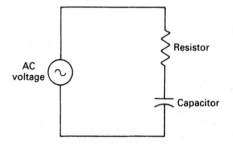

Figure 2–16. Resistor-capacitor series circuit.

[1](*NEC: Article 460; page 486*)

covering of the capacitor in addition to the amount of capacitance and is usually specified within a certain operating temperature range. Ordinarily, if the temperature at which the capacitor is to work is higher than the WV, the working voltage is reduced.

Capacitor Leakage

Since capacitors are constantly subjected to electrical pressure, it is possible for their electrical charge to leak. *Leakage* is the term used when the capacitor manages to self-discharge, thus reducing its effectiveness.

Capacitive Reactance

A capacitor can be used in either dc or ac circuits. When used in an ac circuit, it has a definite opposition to the flow of current. Expressed in a formula, this is stated as

$$X_c = \frac{1}{2\pi f C}$$

Like inductive reactance, capacitive reactance is expressed in ohms. In this formula the Greek letter pi (π) has a value of 3.1416, f is the frequency in hertz (or cycles per second), and C is the capacitance in farads. If the capacitance value is supplied in microfarads or picofarads, these must first be converted to farads prior to using the formula.

Capacitors and Phase

Capacitors and coils behave in an opposite manner as far as phase is concerned. For a coil, the current flowing through it lags the voltage across it by a maximum of 90°. For a capacitor, the current flowing into a capacitor leads the voltage across it by a maximum of 90°. A coil is sometimes described as having a *lagging phase* and a capacitor as having a *leading phase*.

Chapter 3

Wire[1]

Electrical energy is delivered by wires starting at the power plant of an electric utility and supplied to a service box in a home, factory, business, farm, or manufacturing plant. Thus there is a physical connection via wires between the utility and that service box. At the service box, the incoming voltage is routed to a number of branch lines, and so those lines represent a distribution network.

In-home wiring and power distribution are not the responsibility of the utility but that of the owner of the property using the power. While the utility represents what seems to be an inexhaustible source of power, there is always a question as to whether the wiring following the service box can accommodate increased demands for electrical power. Such demand takes place the moment appliances are plugged into receptacles and turned on. It is possible to overload wiring, just as it is possible to overwork or overload an appliance.

CONDUCTORS[2]

A conductor is any substance, gas, liquid, or solid, that permits the passage of an electrical current. Whether a conductor permits an easy passage is another matter. For branch wiring, the most common conductor is copper (*NEC: Pages 26, 27*).

[1]*(NEC®: Article 760–27, page 836)*
[2]*(NEC: Article 100; page 26)*

Copper wiring is preferable since volume for volume it has better conductivity than most other substances. Silver is a notable exception, but its cost makes its application as a branch conductor prohibitive. However, silver is used as a coating on some electrical components that require high conductivity and in some applications in specialty wiring. Gold is not as good a conductor as either silver or copper, but it is used as a plating substance where oxidation should be avoided. The chart in Table 3–1 shows the relative conductance of various substances using silver as the reference. Carbon is a very poor conductor and yet it is used in some microphones.

AMPACITY[3]

Ampacity is a word derived from "amperes" and "capacity." It is the amount of current, measured in amperes, that a conductor can carry without exceeding its temperature rating. The ampacity of a wire is determined by a number of factors: gauge, type of insulation, wire temperature, ambient temperature, material of which the wire is made, and conditions of use.

The chart in Table 3–2 supplies data about the current-carrying capacity of some copper wires. The smaller-gauge wires, such as Nos. 1, 2 and 3, are supplied for comparison purposes. (*NEC: Article 310–15; page 190*)

TABLE 3–1. RELATIVE CONDUCTIVITY OF VARIOUS SUBSTANCES

Substance	Relative Conductance (silver = 100%)
Silver	100
Copper	98
Gold	78
Aluminum	61
Tungsten	32
Zinc	30
Platinum	17
Iron	16
Lead	15
Tin	9
Nickel	7
Mercury	1
Carbon	0.05

[3](*NEC: Table 310–16; page 191*)

TABLE 3–2. APPROXIMATE CURRENT RATINGS

Gauge	Amperes
14	15
12	20
10	30
8	40
6	55
4	70
3	80
2	95
1	110

WIRE SIZE[4]

Wire size is determined by its cross-sectional area and not by its length. The area is a fixed amount; the length is variable. The cross-sectional area is expressed in circular mils (CM). The circular mil is a basic unit and represents the area of a wire whose diameter is 1 mil. One mil is equivalent to $\frac{1}{1000}$ inch or 0.001 inch (Figure 3–1).

Wire size is specified by numbers, ranging from 0000, the thickest wire, to 40, the thinnest. Thus, as the diameter of a wire decreases, its gauge number increases.

Table 3–3 is a listing of some of the physical characteristics of standard annealed solid copper wire, with the listing designated as American Wire Gauge (AWG). The first column in this table indicates the gauge number; the next column is the diameter of the wire in mils. The cross-sectional area of the wire is supplied in circular mils and also in square inches. The area in circular mils is obtained by squaring the diameter, that is, multiplying it by itself. This does not supply a precise value, but rather a reasonable approximation.

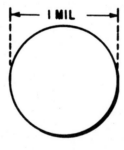

Figure 3–1. The diameter of wire is measured in mils. To determine the area in circular mils, multiply the diameter by itself.

[4](*NEC: Article 760–27; page 836*)

TABLE 3–3. AMERICAN WIRE GAUGE (B & S)

Gauge number	Diameter (mils)	Cross section		Ohms per 1000 feet	
		Circular mils	Square inches	25°C (= 77°F)	65°C (= 149°F)
0000	460.0	212,000.0	0.166	0.0500	0.0577
000	440.0	168,000.0	0.132	0.0630	0.0727
00	365.0	133,000.0	0.105	0.0795	0.0917
0	325.0	106,000.0	0.0829	0.100	0.116
1	289.0	83,700.0	0.0657	0.126	0.146
2	258.0	66,400.0	0.0521	0.159	0.184
3	229.0	52,600.0	0.0413	0.21	0.232
4	204.0	41,700.0	0.0328	0.253	0.292
5	182.0	33,100.0	0.0260	0.319	0.369
6	162.0	26,300.0	0.0206	0.403	0.465
7	144.0	20,800.0	0.0164	0.508	0.586
8	128.0	16,500.0	0.0130	0.641	0.739
9	114.0	13,100.0	0.0103	0.808	0.932
10	102.0	10,400.0	0.00815	1.02	1.18
11	91.0	8,230.0	0.00647	1.28	1.48
12	81.0	6,530.0	0.00513	1.62	1.87
13	72.0	5,180.0	0.00407	2.04	2.36
14	64.0	4,110.0	0.00323	2.58	2.97
15	57.0	3,260.0	0.00256	3.25	3.75
16	51.0	2,580.0	0.00203	4.09	4.73
17	45.0	2,050.0	0.00161	5.16	5.96
18	40.0	1,620.0	0.00128	6.51	7.51
19	36.0	1,290.0	0.00101	8.21	9.48
20	32.0	1,020.0	0.000802	10.4	11.9
21	28.5	810.0	0.000636	13.1	15.1
22	25.3	642.0	0.000505	16.5	19.0
23	22.6	509.0	0.000400	20.8	24.0
24	20.1	404.0	0.000317	26.2	30.2
25	17.9	320.0	0.000252	33.0	38.1
26	15.9	254.0	0.000200	41.6	48.0
27	14.2	202.0	0.000158	52.5	60.6
28	12.6	160.0	0.000126	66.2	76.4
29	11.3	127.0	0.0000995	83.4	96.3
30	10.0	101.0	0.0000789	105.0	121.0
31	8.9	79.7	0.0000626	133.0	153.0
32	8.0	63.2	0.0000496	167.0	193.0
33	7.1	50.1	0.0000394	211.0	243.0
34	6.3	39.8	0.0000312	266.0	307.0
35	5.6	31.5	0.0000248	335.0	387.0
36	5.0	25.0	0.0000196	423.0	488.0
37	4.5	19.8	0.0000156	533.0	616.0
38	4.0	15.7	0.0000123	673.0	776.0
39	3.5	12.5	0.0000098	848.0	979.0
40	3.1	9.9	0.0000078	1,070.0	1,230.0

Standard annealed solid copper-wire table using American Wire Gauge (AWG). Another gauge, Browne & Sharp (B & S), is identical to AWG.

Columns 5 and 6 supply the resistance of the wire in ohms per 1000 feet. The resistance is measured at 25° and 65° Celsius. Copper has a positive temperature coefficient of resistance; that is, its resistance increases as the temperature rises.

For the most part, the gauge numbers supplied in this table are not of interest to many electricians. Number 14 gauge is quite common. Gauges 16 and 18 are used in appliance cords. Service entry wires are commonly No. 6 or 8. Fine wires, gauge No. 30 and higher, are used in some manufacturing processes.

As the cross-sectional area of a wire is increased, that is, as the gauge number is decreased, the greater is its ampacity, the ability to carry an electric current with minimum heat loss. The actual current capacity of a wire depends not only on its cross-sectional area, but also on both the ambient and wire temperatures, the length of the conductor, the insulation surrounding the wire, and the time duration of usage. Another factor is the ability of the wire to radiate its heat into open space.

Voltage Drop in a Feeder Line[5]

Appliances are not only rated in terms of power they use but also in the amount of voltage they must have to operate properly. To calculate the voltage drop along a feeder line:

$$E = I \times \frac{K \times L}{CM}$$

In this formula, E is the amount of the voltage drop, I is the current in amperes, for copper, K is a constant and has a value of 10.8, L is the one-way length of the copper wire in feet, and CM is the area of the conductor in circular mils. As a start, it will be necessary to assume the wire gauge and then to obtain the cross-sectional area in circular mils from the wire table, Table 3–3.

Assume this wire will be No. 10 gauge. The table indicates that the area of this wire will be 10,400 circular mils. If the distance from the line voltage source to the machinery is 50 feet then the voltage drop with a current of 5 amperes will be:

$$E = 5 \times \frac{10.8 \times 50}{10,400} = 0.26 \text{ volt}$$

This is a very small amount. If the line voltage at the source is 120 volts, then the voltage at the equipment will be 120 − 0.26 or 119.7 volts.

Whether this drop is significant or not depends on the operating voltage

[5](NEC: Article 210–19; page 58)

range of the equipment. Larger amounts of current and a higher wire-gauge number may make the voltage drop significant.

Temperature

Operational values of electrical components and devices are often specified with reference to temperature. An electrician may be interested in knowing the final, working temperature of a motor or generator armature, of a particular kind of lamp, or the value of resistance at a certain temperature. In electrical work temperature is generally specified in degrees Fahrenheit or Celsius (formerly called Centigrade). There are other temperature scales, such as Kelvin, Reaumur, and Rankine, but these are more often found in laboratory rather than commercial or industrial use.

To convert degrees Celsius to degrees Fahrenheit:

$$F° = (C° \times \tfrac{9}{5}) + 32$$

To convert degrees Fahrenheit to degrees Celsius:

$$C° = (F° - 32) \times \tfrac{5}{9}$$

Positive temperature coefficient.

Positive and negative numbers are used in electrical work. Consider any electrical conductor, such as a length of copper wire or the filament of a light bulb. These conductors, like all others, have a certain amount of resistance per unit length. This resistance is not constant however, but changes with temperature. Therefore, a conductor has not one, but two resistances. One is the cold resistance; the other the hot resistance. For the filament of a lamp or copper wire, the resistance rises with increases in temperature. Described another way the copper wire or lamp filament has a positive temperature coefficient of resistance.

Since temperature does have an effect on resistance, the resistance of wire is often specified at a particular temperature. Commonly indicated temperatures are 25° Celsius (77° Fahrenheit) and 65° Celsius (149° Fahrenheit). For example a certain wire might have a resistance of 1.02 ohms per 1000 feet at 25° C with an increase to 1.18 ohms per 1000 feet at 65° C.

Is such a small increase in resistance important? It is when working with long lengths of wire and heavy currents. It is if you consider an increase in resistance requires a corresponding increase in voltage to maintain the current at a desired level.

Temperature coefficient of resistance.

Different types of conductors do not increase equally in resistance with temperature since resistance depends on the substances of which the conductors are made. For all conductors such as copper, aluminum, or silver, each degree rise in temperature is accompanied by a constant increase in resistance. This constant increase in resistance is called the temperature coefficient of resistance. The temperature coefficient of resistance is the amount of increase of resistance per ohm per degree rise in temperature above 0° Celsius. For copper, the temperature coefficient of resistance is 0.00427 ohm. Thus, if you have a wire that has a resistance of 1 ohm at 0 C, it will have a resistance of 1 + 0.00427 ohms at 1° C. The temperature has risen by 1 and the resistance of the wire has increased by 0.00427 ohm.

Be careful. The resistance increase of 0.00427 ohm is for each ohm of resistance of the conductor. If you start with a wire of 2 ohms resistance at 0° C and increase the temperature to 1° C, then the increase in resistance will be 2×0.00427. The final resistance will be 2 ohms (the original resistance) + $2 \times 0.00427 = 0.00854$ ohm or 2.00854 ohms.

Resistance of Wire[6]

The resistance of wire per unit length, columns 5 and 6 in Table 3–3, depends on the cross-sectional area. When using wire leading from the fuse or circuit breaker box, the gauge to select depends on the load, plus a safety factor, and whether the load will increase in the future.

Effects of Inadequate Wire Gauge

There are two possibilities when using a larger than required gauge of wire for current-demanding appliances. Because of the greater resistance of the wire, there may be an unacceptably high voltage loss (a voltage drop along the wire), and so the appliance (or appliances) may not receive an adequate amount of input voltage. This may affect appliance performance. Even more objectionable is that the wire may become exceptionally warm or hot. Because of the insulation around the wire and the closed-in condition of the wire, the heat will not be able to dissipate and can result in a fire.

If you know the resistance of 1000 feet of any wire, it is possible to calculate the resistance of 1000 feet of wire of the next larger gauge number. Thus, the resistance of 1000 feet of gauge 20 wire at 25° C is 10.4 ohms. Multiplying this by 1.26 supplies a resistance of 13.1 ohms for 21-gauge wire. Conversely, to find the resistance of 1000 feet of the next lower gauge number, divide by 1.26. The resistance of smaller lengths can be calculated by multiplying or dividing by

[6](*NEC: Chapter 9, Table 8; page 888*)

some factor such as 10 or 100. Thus, the resistance of 300 feet of 30-gauge stranded wire is $(300/1000) \times 105.0 = 31.5$ ohms. The 105.0 ohms is the resistance of 1000 feet of this wire. The resistance of a wire is proportional to its length in feet and inversely proportional to its cross-sectional area.

Table 3–3 does not include the weight of the wire. The data are for the wire itself and do not include the insulation, since its weight is a variable and depends on the insulation material and the amount that is used. On average, 1000 feet of No. 40 wire weighs 0.0299 pounds and 1000 feet of No. 0000 wire weighs 641 pounds.

INSULATION[7]

Insulation is a nonconductive material, and its purpose is to confine the flow of current to a wire. The insulation can be a paper wrap, cotton, silk, plastic, asbestos, neoprene (a type of synthetic rubber), varnished cambric, or any other commercially made substance such as polyvinyl chloride (PVC). It can include enamel, which is fairly easy to scrape away from the wire, or a type of enamel substitute that is almost impossible to remove. In some instances, several layers of the same or different insulating materials are used. The amount of insulation (Figure 3–2) is governed by the amount of applied voltage, the amount of current, the temperature, working conditions, including the possible presence of water vapor, and the need for confining the current to the conductor.

Insulation can be an integral part of a wire or it can be applied by an electrician. At one time, rubberized tape was used, but this had a tendency to unravel. The preferred form is plasticized electrical tape.

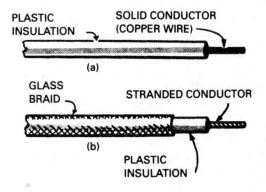

Figure 3–2. (a) Wire with single layer of insulation; (b) with double layers of insulation.

[7](NEC: Article 310–12; page 181)

WIRE TYPES[8]

There are various ways of designating wire. Two possible listings would be solid and stranded. Another would be indoor and outdoor wire. Still another method would be by the kind of insulation used.

Stranded and Solid Wire[9]

Both types of wire, stranded and solid, have advantages and disadvantages. Flexibility is an advantage of stranded wire and in this respect it is easier to handle than solid. However, when stripping insulation from stranded wire, it is very easy to cut away and lose one or more strands. Solid wire is not as flexible, but it is easier to remove its insulation. Solid wire is easier to splice or tap. As a general rule, wires having about a No. 30 or 34 gauge are stranded (*NEC: Article 310–6; page 178*).

The cross-sectional area of No. 8 wire is about 10 times as great as that of No. 18 wire. But as a wire is made thicker, it becomes increasingly difficult to handle. It is quite easy to wrap a No. 16 wire around the screw terminal of a switch or a receptacle. Doing it with a No. 8 wire is quite an accomplishment.

Stranded wire consists of wires twisted together. The strands, if made of copper, are primarily covered by a coating of natural rubber or some synthetic. This first coating can be covered with one or more wraps of braid. When connecting stranded conductor to another wire or to a screw terminal, be careful not to break any of the wire strands. Twist them to ensure good electrical contact.

Identifying Single or Stranded Wire[10]

When buying wire, if a single number is indicated, then the wire is a solid type. Thus, the label on a spool of wire will state No. 18, or some other number. Stranded wire is always referred to by a pair of numbers, such as 18/30. This means that the conductor consists of 18 strands of No. 30 wire.

Current Capabilities of Stranded Wire

Like solid wires, stranded types have current limitations. Sixty-five strands of No. 34 wire will carry the same amount of current as a solid No. 16 wire. Instead of referring to the wire as 65/34, some manufacturers will simply identify it as No.

[8](*NEC: Article 318–12; page 225*)
[9](*NEC: Article 800–2; page 852*)
[10](*NEC: Article 310–3; page 176*)

16 stranded. Many combinations of stranded wires could be made into an equivalent of No. 16. The thicker the wire, the fewer the number of strands. And so a designation such as No. 16 stranded does not tell you anything about the number of strands nor the gauge of the individual wires.

How to Calculate the Solid Conductor Equivalent of Stranded Wire

Assume you have stranded wire (Figure 3–3) consisting of 37 strands, each of which is 2 mils (0.002 inch thick). To calculate the area in circular mils, multiply the diameter in mils by itself. $2 \times 2 = 4$, so each individual strand has a crosssectional area of 4 circular mils. However, there are 37 strands, and so the total cross-sectional area is 37 times $4 = 148$ circular mils.

Consult Table 3–3; under the heading "Cross Section" and in the third column you will find areas in circular mils. Move down this column and you will see that there is no single solid conductor that has an area of 148 circular mils. Number 29 wire is 127 circular mils and No. 28 wire is 160 circular mils. So 37 strands of 2-mil wire could be considered somewhere between Nos. 28 and 29 gauge solid conductor, but much closer to No. 28 than to No. 29.

The chart in Table 3–4 supplies data for stranded wires ranging from Nos. 32 to 10. Thus, seven strands of 3.1-mil diameter supplies the equivalent of No. 32 AWG solid conductor.

Wire gauges such as No. 6 or even No. 8 are too rigid to be pulled through conduit. The stranding of wire follows an established standard. The total number of strands and the size of each are listed in Table 8, Chapter 9, page 888 of the NEC.

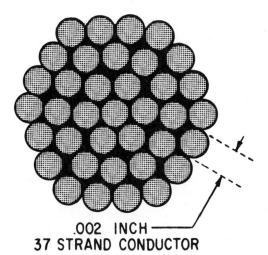

.002 INCH
37 STRAND CONDUCTOR

Figure 3–3. Cross-sectional view of 37 strands of 2-mil wire.

TABLE 3–4.   DATA FOR STRANDED WIRE

Size approx. (AWG)	Number of strands	Uninsulated conductor			Finished wire
		Strand diameter, nominal (inches)	Strand area (circular mils)	Condcutor diameter, average (inches)	DC resistance (ohms/1000 ft)
32	7	0.0031	67	0.010	183.0
30	7	0.0040	112	0.013	109.9
28	7	0.0050	175	0.016	70.4
26	1	0.0159	253		46.3
26	7	0.0063	278	0.020	44.3
24	1	0.0201	404		28.3
24	7	0.0080	448	0.025	27.5
24	19	0.0050	475	0.026	25.7
22	1	0.0253	640		17.9
22	7	0.0100	700	0.031	17.6
22	19	0.0063	754	0.032	16.3
20	1	0.0320	1024		11.2
20	7	0.0126	1111	0.038	10.9
20	10	0.0100	1000	0.038	12.3
20	19	0.0080	1216	0.041	10.1
18	1	0.0403	1624		7.05
18	7	0.0159	1770	0.048	6.89
18	16	0.0100	1600	0.048	7.69
18	19	0.0100	1900	0.051	6.48
16	1	0.0508	2581		4.43
16	19	0.0113	2426	0.058	5.02
16	26	0.0100	2600	0.061	4.73
14	1	0.0641	4109		2.79
14	19	0.0142	3831	0.072	3.18
14	41	0.0100	4100	0.076	3.00
12	1	0.0808	6529		1.76
12	19	0.0179	6088	0.090	2.00
12	65	0.0100	6500	0.096	1.89
10	104	0.0100	10380	0.121	1.16

BASIC TYPES

There are a number of basic wire types. These include bare solid copper, bare stranded wire (Figure 3–4), insulated wire, and metallic shielded wire. There are numerous variations under these four types, with some wire made for specific applications such as audio and television cable. Wire can be single or a parallel pair or a twisted pair. It is available as a single conductor surrounded by a metallic

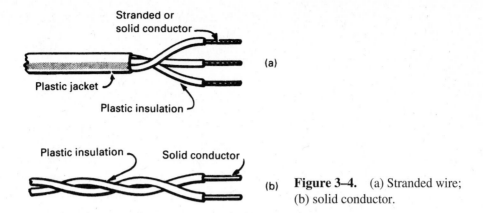

Figure 3–4. (a) Stranded wire; (b) solid conductor.

shield made of a braid that also works as a conductor. It can be made for indoor use only and for outdoor use, both above and below ground.

Single Conductor

The simplest type of wire is bare, solid conductor. It has application as a grounding wire in wiring installations.

Although single wire can be bare, as in the case of a ground wire, it can also be insulated with various types of coverings. One of these is enamel, but it has the disadvantage that it can flake or be easily removed. Another type of single wire is tinned with a thin layer of tin or solder. Its advantage is that it can be soldered very easily. Still another single wire is known as Bell wire or pushback wire, so called since its insulation can easily be pushed back to expose the bare copper wire beneath. At one time it was known as annunciator wire because of its use in electrical signaling devices. Bell wire is insulated with a double layer of cotton, with each layer wound in opposite directions and impregnated with paraffin (Figure 3–5).

Still another type of single wire is referred to as single-wire line. It differs from Bell wire since it is covered with plastic, which is not as easily removed as the cotton covering of Bell wire. The plastic insulation may have no color, but it is also available in various colors. The insulation is removable with a wire stripper or with a single-edge razor blade.

Figure 3–5. Bell wire. Designed for low-voltage use only. Insulation is easily pushed back for making connections.

Sometimes insulated single conductors are wrapped around each other, forming a twisted pair.

BX[11]

BX (Figure 3–6) is flexible armored cable containing two, three, or four conductors. At one time, BX was widely popular for use in in-home wiring, but it has largely been replaced by Romex. Ordinary BX is not suitable for outdoor use. BX is usually found in older homes, and in some locations it does not meet with the approval of municipal electrical authorities. The advantage of BX is that it is a metallic shield complete with its wires, and so snaking the wires through the shield is not necessary. However, this is not as much of an advantage as it might seem, since in cutting away the outer metallic sheath to get at the wires it is possible to cut through the insulation and damage the wires. Also, the rough edges of the BX that have been cut can damage the insulation unless it is protected by a bushing.

The code name for BX as designated in the National Electrical Code is type AC or ACT. The wires are covered with insulation and this has successive wraps of strong paper. The purpose of the paper is to protect the insulation from the inner wall of the metallic shielding. Between this paper wrap and the inner wall, there is a continuous wire that is used for the ground connection. The outer metal shield also acts as a ground, but the shield should not be depended on alone for grounding.

BX can be regarded as intermediate between Romex and conduit. It is much easier to make an in-wall installation using Romex, but BX can easily be used in attics and basements having exposed beams.

Earlier BX cable, still used in older homes, is a two-wire type consisting of a black (hot) lead and a white (ground lead). Subsequently, the two-wire type was replaced by three-wire BX that used a black and white lead, plus an unshielded

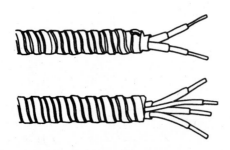

Figure 3–6. Two-wire and four-wire BX.

[11](*NEC: Article 333; page 242*)

ground lead. While BX is still available, it is not used in new home construction and in some areas is a violation of municipal electrical codes.

BX is difficult to handle, but it is still easier than rigid wall conduit. It is relatively flexible, and so it is possible to bend it without the use of a bending tool. Always make sure, when buying BX, that it has a third wire, a ground wire, running through it. This wire is usually bare.

BX cannot be squared off; that is, it is not possible to make right-angle bends. Make the bend as gradual as possible, with the turn radius about five times the cable diameter (*NEC: Article 333–8; page 243*). Sharp turns will damage the armor and cause adjacent turns to separate. Once this happens, there is no way to get the armor back into its original position.

Supporting BX.[12] If the wall studs are available and you are able to drill holes through them, you can use them to support the cable (Figure 3–7). Make the holes at least $\frac{1}{8}$ inch in diameter larger than the diameter of the BX so that the cable can pass through without difficulty.

If the BX is to run along a beam, use staples of the kind shown in Figure 3–8. The customary technique is to support the BX with staples every 4 to 5 inches. If the BX shows a tendency to sag, decrease this distance. When bringing the cable into an electric box, make sure the BX has some support not more than 2 to 3 inches away. The reason for this is to avoid any mechanical strain on splices that are to be located inside the box.

Strain relief. No current-carrying wire should have any strain put on it; that is, it must not be used to support any object, not even anything as light as a plug. A wire entering a plug has a strain relief in the form of an Underwriters' knot. A wire entering an electric box may not be used in any way to support that

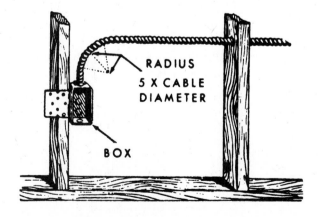

RADIUS
5 X CABLE
DIAMETER

BOX

Figure 3–7. Stud used for BX support.

[12](*NEC: Article 333–7; page 243*)

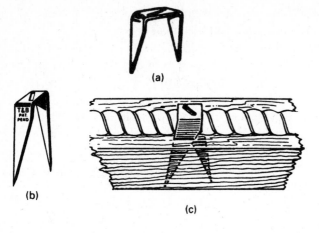

Figure 3–8. Support staples for BX. (a) Straight staple and (b) off-set type. (c) Phantom view shows how an offset staple holds and supports BX.

box. External to the box, wires are made self-supporting by staples. Following its entry into an electric box, either BX or Romex must be supported by a clamp or a fitting.

Cutting BX.   Prior to cutting BX, carefully measure the distance from the starting electric box to the finish box. Don't make the cable too short and make allowances for all bends. If, during installation, the cable is too short, do not splice and add on more cable. Instead, cut a completely new section and save the spoiled section for some subsequent, shorter run.

Working with BX.   The best tool for cutting BX is a special BX cutter consisting of rotary blades made to turn against the outer surface of the metal shell. It is easy to use and is desirable for trimming the metal shield away from the cable. It is also possible to use a hacksaw, but it requires much more care and more work.

When using a hacksaw, always start with a new blade. Select a quality blade intended for metal cutting. When inserting the blade into the hacksaw frame, make sure the teeth point away from the handle. Also make sure the blade is tight and that it does not bend or curve. Hacksaw blades can snap if they are not inserted or used properly.

Don't try to cut through the BX by holding it. Use a vise. This is a preparatory step to cutting the BX to its correct length.

After this preliminary step is completed, the next step is to remove enough of the armor to expose the wires. It will be necessary to cut very carefully to avoid damaging the enclosed wires. Cut the armor about 8 inches from the end so that there will be enough wire to go into an electric box. Put the BX in a vise with an 8-inch length of the cable extending beyond the end of the vise. Cut at right angles to the cable, with the hacksaw blade up against the vise. In this way the vise will act as a cutting guide.

Do not let the hacksaw blade cut continually at one spot, but move it to cut a semicircle. Make each stroke light and do not exert too much pressure. When the blade begins to cut through the armor, it will then tend to grip the blade, an indication that a stop point has been reached. Loosen the vise, turn the cable, and continue cutting, again in semicircular form.

Upon completion of these steps, there will be a cut circle around the BX. Keep deepening the cut, being careful not to cut into the surface of any of the enclosed conductors. Finally, the cut will be so deep that it will be possible to break the armor by flexing the cable back and forth (Figure 3–9). If some of the armor remains unbroken, cut it away with the hacksaw. Be careful. The edges of the cable can cut into a finger.

It should now be possible to pull the two sections away from each other, exposing the wires. Undo the paper wrapping around the wires and cut it away with scissors.

One end of the BX has now been completed. Repeat the process at the other end. The cut edges at both ends of the BX can be rough. File them with a flat file, but once again be careful not to damage the insulation around the wires.

The edges of the cut around the BX may still be sharp. Slide an antishorting bushing at each end and then insert each bushing so it fits snugly between the wires and the inside of the BX armor (*NEC: Articles 333–9, 333–11; pages 243, 244*).

Romex[13]

Romex is a nonmetallic, plastic-covered cable now widely used for wiring and is available with or without a ground wire; that is, it can be either two- or three-wire cable. It is easier to handle than BX.

The NEC requires that all electrical conductors be imprinted with data defining the AWG size, the type of insulation (specified by a letter code), the

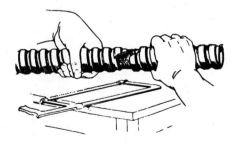

Figure 3–9.   After cutting through the armor, rotate the two sections until they break away. The wires and a paper wrapping around them will now be exposed.

[13](*NEC; Article 336c; page 250*)

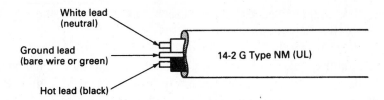

Figure 3–10. Coding on Romex. 14 refers to the wire gauge. 2G is the number of conductors, not including the ground wire. UL is an abbreviation for Underwriters' Laboratories. NM is nonmetallic sheathed cable.

number of conductors, and a UL designation. A typical wire imprint is illustrated in Figure 3–10.

It is practically impossible to visually identify most cables, since they look so much alike, consequently, they carry data throughout the length of the external wrap. Figure 3–10 shows how this is done. The number 14 in this case indicates the wire gauge, and the following number 2 means the cable consists of two wires, one of which (the black lead) is to be the hot wire, and the other (the white lead) is to be the neutral. The letter G means that the cable carries a separate ground wire. NM is a reminder that the cable is to be used only in a dry location. A dry location means one that is always dry, and not practically dry or dry only from time to time.

Wire Color Code

The color coding of wires depends on their function. Black is the most commonly used color for a hot lead. If there is more than one hot lead, red is used for identification. A third hot lead would be coded blue, and a fourth would be yellow.

The most common color for a neutral wire is white. Less often, gray is used.

The ground lead is often a bare wire. If insulated, it is color coded green or it may be green with yellow stripes. The color coding is used on the insulation surrounding the conductor.

Conductor identification.[14]   Wire can be identified by its gauge number (that is, its thickness), by whether it is bare or insulated, by the type of insulation, by whether it is single or in cable form, by number, by whether it is low-voltage or house-voltage wire, or by its intended use. Its possible function can include lamp cord, extension cord, in-wall wiring, flat cable, round cable, and indoor or outdoor use. Wires are often also identified by letters. A requirement of

[14]*(NEC: Article 310; page 176)*

the National Electrical Code is that all wire used in building construction be marked with its gauge number along its length.

Insulation Code

Wire must carry a letter code indicating the type of insulation. The following listing summarizes the most commonly used letter codes.

NM	Nonmetallic cable for indoor use only. Better known as Romex. Used in dry locations only.
NMC	Nonmetallic cable for either indoor or outdoor use. The protective covering is tougher than that of NM, although this is not evident by purely visual inspection. Used in dry or wet locations.
UF	Underground fused non-metallic cable. Can be substituted for NMC.
USE	Non-metallic underground service entrance.
MV	Medium voltage cable. Cable that is rated above 2,000 volts.
FCC	Flat conductor cable. For use with branch circuits under carpet squares; for floors that are hard and made of materials such as concrete, cement and wood.
SPT	Lamp cord.
HPN, HPD	Heater cord.
RH	Rubber insulated wire. R is rubber insulated wire that is not as good quality as RH.
T, TW	Temperature. Thermoplastic insulation for in-home wiring.
W	Wet
SV, SJ	Extension cords, SV is designed only for vacuum cleaners.

WIRE LOCATIONS[15]

Wires can be installed in three types of locations: damp, wet, and dry. These locations are defined in the Code, which should be consulted for a guide to the type of wire to be used. Since it can be difficult to determine whether a particular location is damp or wet, the safest procedure to follow is the NEC recommendation for a wet location.

[15]*(NEC; Article 310–8; page 179)*

LENGTHS OF WIRE AND CURRENT RATING OF EQUIPMENT

The greater the maximum possible current demand of a load and the greater the distance of that load from the service box, the smaller is the required wire gauge. Table 3–5 shows the relationship between current and wire lengths in feet. The data supplied in Table 3–5 are approximations. The NEC provides a number of tables in Chapter 9 plus specific examples for various types of conductors.

CONDUIT[16]

Conduit is a hollow pipe used for enclosing wires and can consist of two basic types, rigid and flexible, although there are a number of variations. The conduit can be metallic or nonmetallic. Conduit can also be categorized as plastic, thin-wall metal, and threaded. The purpose of conduit is to protect the wires it encloses. The advantage of Romex over conduit-enclosed wire is that Romex can be snaked through walls. The diameter for the opening of conduit can range from $\frac{1}{2}$ inch to 3 inches.

Rigid Metal Conduit[17]

Rigid conduit has the same size designations as water pipe. The size of the conduit is specified by its inside diameter (ID). One-half-inch conduit has an actual ID of 0.622 inch. Standard conduit sizes used in interior wiring range from $\frac{1}{2}$ inch to 3 inches, but larger sizes are available for commercial installations (*NEC: Article 346–15a; page 267*).

TABLE 3–5.   WIRE SIZE AND CURRENT RATING OF EQUIPMENT

Rating of equipment (A)	Use no. 18 for lengths (ft):	Use no. 16 for lengths (ft):	Use no. 14 for lengths (ft):	Use no. 12 for lengths (ft):
5.8	1–125	126–150	151–190	191–295
7.2	1–50	51–150	151–180	181–230
9.8	1–50	51–125	126–150	151–180
13.0		1–50	51–125	126–165
15.0			51–150	51–150
20.0				1–50

[16](*NEC; Chapter 9, Table 1, Page 879*)
[17](*NEC: Article 100; page 26*)

Wires for Rigid Conduit

There are some wires whose outer insulation is especially prepared to facilitate fishing through rigid conduit. This treatment consists in giving the insulation a wax coating so that the wires can slide more easily against each other and the inner wall of the conduit. Thermoplastic insulation types (T and TW) seem to be preferred.

The metal used for making this type of conduit is either steel or aluminum. It can be used indoors or outdoors and is the type that is preferred for underground work, but it is advisable to check with municipal electrical authorities to learn if it may be used underground. Underground rigid conduit may be permitted, but there may be some restrictions on installation methods, since some soils may be too corrosive.

Rigid metal conduit is sometimes confused with water pipe, which it resembles. The interior and the exterior may be galvanized, and in some conduit there may be an additional coating of enamel or plastic. It is essential for the interior to be smooth so as not to snag or abrade the insulation of the wires being fished through. It is possible to buy rigid conduit cut to length and threaded or to use a tool for cutting threads. The usual length for conduit is 10 feet, with each length carrying a UL label.

Galvanized conduit is generally approved for indoor or outdoor installation, while the black-enamel type is restricted to indoor use. Rigid conduit costs more than other types, such as NM cable, but it can carry more conductors in a single run. With BX the wire is an integral part, and so there is no option for adding more wires.

Conduit Diameter and Wire Gauge[18]

The number of wires that are permitted in conduit is determined by two factors: the gauge of the wire and the diameter of the conduit. The relevant data are listed in Table 3–6.

Conduit Designations

Various abbreviations are used for conduit, most commonly EMT (electrical metallic tubing), PVC (rigid nonmetallic), ENT (electrical nonmetallic tubing), and IMC (intermediate metallic conduit).

EMT.   One of the more popular forms of conduit is EMT. It has a thinner wall than regular conduit, but its interior diameter and cross-sectional area are the

[18](*NEC; Article 370–16; page 303*)

TABLE 3–6. CONDUIT DIAMETER VERSUS WIRE GAUGE

Avg. size wire	Minimum size conduit permitted (in).							
	1	2	3	4	5	6	7	8
14	$\frac{1}{2}$	$\frac{1}{2}$	$\frac{1}{2}$	$\frac{1}{2}$	$\frac{3}{4}$	$\frac{3}{4}$	1	1
12	$\frac{1}{2}$	$\frac{1}{2}$	$\frac{1}{2}$	$\frac{3}{4}$	$\frac{3}{4}$	1	1	1
10	$\frac{1}{2}$	$\frac{3}{4}$	$\frac{3}{4}$	$\frac{3}{4}$	1	1	1	$1\frac{1}{4}$
8	$\frac{1}{2}$	$\frac{3}{4}$	$\frac{3}{4}$	1	$1\frac{1}{4}$	$1\frac{1}{4}$	$1\frac{1}{4}$	$1\frac{1}{2}$
6	$\frac{1}{2}$	1	1	$1\frac{1}{4}$	$1\frac{1}{2}$	$1\frac{1}{2}$	2	2
4	$\frac{1}{2}$	$1\frac{1}{4}$	$1\frac{1}{4}$	$1\frac{1}{2}$	$1\frac{1}{2}$	2	2	2
2	$\frac{3}{4}$	$1\frac{1}{4}$	$1\frac{1}{4}$	2	2	2	$2\frac{1}{2}$	$2\frac{1}{2}$
1	$\frac{3}{4}$	$1\frac{1}{2}$	$1\frac{1}{2}$	2	$2\frac{1}{2}$	$2\frac{1}{2}$	$2\frac{1}{2}$	3
0	1	$1\frac{1}{2}$	2	2	$2\frac{1}{2}$	$2\frac{1}{2}$	3	3
00	1	2	2	$2\frac{1}{2}$	$2\frac{1}{2}$	3	3	3

same. EMT is available in sizes from $\frac{3}{8}$" to 2". One of its advantages is that the inside surface is enameled. This supplies a smoother inside surface making it easier to fish wires through it. Enamel is also an insulator so, presumably, there may be some added protection in the event one of the conductors happens to make contact with the inner wall. However enamel scrapes and flakes easily.

Outdoor Conduit

Selected types of conduit are suitable for outdoor use. These include aluminum, steel, and plastic types such as polyvinyl chloride (PVC), or high-density polyethylene. Which type to use depends on whether the conduit is for above ground applications or underground. For above ground use PVC; for underground use polyethylene. PVC reacts chemically with sunshine and to prevent changes in its composition it can be coated with outdoor latex paint. Make certain to cover the conduit completely by using two coats.

The inside diameter of the conduit must be wide enough to permit the use of AWG No. 12/3 UF cable (3 strands of No. 12 AWG wire). Use $\frac{3}{4}$-inch conduit.

Greenfield

There is still another type of conduit known as Greenfield now largely replaced by plastic types. Greenfield resembles BX but is supplied without wires so these must be fished through. Greenfield is metal and its structure is such that it is quite flexible. It can be cut with the same type of cutter used for BX or by using a hacksaw. Its ends should be fitted with bushings to prevent possible damage to the wires.

Its flexibility can lead to sagging, so it should be supported within one foot of its entry or exit of an electric box, and also at separations between four and five feet. Its flexibility also means it has a higher dc resistance, more so than continuous metallic conduit. For this reason it should include a separate ground wire in addition to any other conductors fished through it.

Conduit Fittings[19]

Conduit can be threaded at the end that fits into an electric box. To fasten the conduit to the box, insert the conduit a short distance into it via a knockout hole and then put a locknut on the threaded end. Put a bushing following the locknut on the conduit. Keep the amount of conduit going into the box at a minimum so as to save as much space inside the box for wiring and for either a switch or receptacle.

The work can be completed by putting a locknut on the conduit leading into the box. Tighten both inside and outside locknuts to hold the conduit firmly in position. Fish the wires through the conduit, but make sure that enough is in the box to permit connections to be made easily. There are various types of locknuts and bushings, but use those whose outer edges are flanged. They are not only easier to tighten but can be tightened more securely.

Selecting conduit is often a matter of personal preference, but one widely used type is a thin-walled, galvanized steel pipe. It supplies better wire protection than either BX or Romex.

Conduit can be extended in two ways: by using an extension coupling or a Condulet. Romex is easiest to use when a right-angle turn needs to be made. BX can make a right-angle turn provided the turn is not sharp. For this purpose, conduit uses an ell joint. If water can enter the conduit use an adhesive on its connecting sections. Once applied the joints cannot be separated.

One type of conduit is made of polyvinyl chloride (PVC) and it can burn. When it does so, it gives off toxic fumes.

Couplings and connectors. If a length of conduit does not have threaded ends, it can still be fastened to an electric box by using a connector. One end is threaded and fits into the box and is held in position for a locknut and bushing. The conduit fits into the other end, outside the box, and is fastened to the connector by either one or two screws. A coupling is used to join two lengths of conduit with the help of a pair of machine screws.

[19](*NEC; Article 342; page 255*)

PLASTIC AND METAL RACEWAYS

In time, just about every home finds that its electrical system is inadequate. It starts by learning that there aren't enough receptacles, or some are not positioned where wanted. This results in a multiplicity of cube taps or using more receptacle strips, thus a branch line can become overloaded. The use of a new branch line is preferable but is the most expensive solution since it can involve behind-the-wall wiring, something not always done easily.

Another solution is to tap into a receptacle that is fairly accessible, but running an outside-the-wall wire isn't attractive. With this problem consider using a raceway. These are slim, rectangular enclosures, made of plastic or metal and can be painted to match the wall covering. Typical connecting wire can be type NM or NMC. To make the connection, use an existing box at the end, in effect adding another receptacle. Use wire gauge No. 12 or No. 14.

Raceways are either one- or two-piece types. The one-piece unit requires fishing wire through. The two-piece unit simply involves laying the wire in place in the raceway, and then covering with the other half of the raceway.

EXTENSION CORDS[20]

There are two basic types of extension cords, indoor and outdoor. They are used for making a temporary connection between a receptacle and some appliance. In effect, they extend the fixed receptacle to a point some distance away and are available in lengths from 6 to 100 feet. For outdoor use, select a cord marked "suitable for outdoor appliances." An indoor cord is not safe for outdoor work.

In its basic form the extension cord consists of a plug at one end and a receptacle at the other. There are variations. The power cord receptacle may be a single- or multiple-access type. It may consist of socket and associated lamp protected by a hinged metal shield. It can also include several immediately adjoining receptacles. The plug for making an electrical connection to a receptacle is usually a three-prong type that supplies automatic grounding.

Like other types of wire, extension cords can be identified by letter, as indicated in Table 3–7. In this table, the top row shows that a 6-foot extension cord can consist of two no. 18 flexible conductors or a pair of conductors made of 41 strands of No. 34 wire and that this cord will have a 7-ampere rating.

[20](*NEC: Article 305–6 (a) and (b); page 175*)

TABLE 3–7. DATA FOR TWO-WIRE EXTENSION CORDS

Conductor Size	Conductor Stranding	Rating (amps)	UL Style	Nominal OD	Cord Length (ft)
18/2	41/34	7	SV	0.245	6
18/2	41/34	7	SV	0.245	9
18/2	16/30	7	SJ	0.300	10
18/2	16/30	7	SJ	0.300	25
16/2	26/30	10	SJ	0.325	10
16/2	26/30	10	SJ	0.325	25
16/2	26/30	10	SJ	0.325	50
16/2	26/30	10	SJ	0.325	100

Selecting the Right Extension Cord

When plugged into a receptacle, an extension cord automatically becomes part of an existing branch power line. The amount of current required by the appliance connected to the cord cannot exceed the current limitations of that branch line. If the extension cord is to be used to operate an appliance that requires relatively little current, there is no problem. Relatively little current means a current of 1 or 2 amperes, or less. This assumes very light loading of other receptacles along the same branch power line. To select the right extension cord, check the amperage rating on the appliance, tool, or equipment that will be used with the cord.

The current taken by the connected tool or appliance must flow through the cord, and so there will be a voltage drop across its length. This reduces the amount of voltage available at the appliance. The greater the amount of current and the lower the current-carrying capability of the cord, the greater will be the voltage drop. It can result in poor operation of the appliance or it may be completely inoperative.

After using the extension cord, remove its plug from its connecting receptacle. Hold the prongs and, if they feel warm, the current capacity of the cord is inadequate for the particular appliance being used, although it may be satisfactory for some other, less current-hungry appliance.

Table 3–7 lists the current-carrying capacities of extension cords based on normal household voltage. The usual voltage drop of a typical cord is about 3% over the length of the cord.

Two-wire extension cords were widely used at one time, but the three-wire type is now more common. The two-wire type has a hot lead and a neutral; the three-wire type also includes a ground wire. The three-wire type is easily recognized by its three-prong plug. Table 3–8 supplies the specifications for three-wire

TABLE 3–8.   GAUGES FOR THREE-WIRE EXTENSION CORDS

	Gauge			
Ampere Rating of Load	25-Foot Cord	50-Foot Cord	75-Foot Cord	100-Foot Cord
1 through 7	18	18	18	18
8 through 10	18	18	16	16
11 or 12	16	16	14	14

extension cords. These cords are available in lengths from 25 to 100 feet, although 'shorter or longer cords can be had. Any length of this cord carries the code name SJ. A 10-foot length contains three No. 18 conductors or is made up of three 16/30 conductors (16 strands of No. 30 wire) and is capable of carrying 7 amperes.

Identifying data on an extension cord can be written in different ways, but all carry the same information. Thus, 16-gauge, three-wire cord may be indicated as 16/3, or 16/3 AWG, or 16 AWG 3, or 3 conductor 16 AWG.

To use extension cords ranging in length from 1 to 295 feet with loads having a 5.8 to 20-ampere rating, consult Table 3–9.

An extension cord is used when the line cord attached to an appliance is too short to reach a power receptacle. Extension cords are also used to bring power to some distant point, such as an electric lawn mower, or to extend the range of a troubleshooting lamp. Extension cords can be as short as 3 feet to more than 100. But whatever the length or usage, an extension cord is just a continuation of the wiring of a branch circuit. This means it must have at least the same wire gauge as the branch or a lower gauge (that is, a thicker wire).

Extension cords have two basic ratings, length and wire gauge, and both of these are marked on the package. Extension cords are identified as medium, heavy, and super, although other equivalent designations may be used.

TABLE 3–9.   LOAD RATING VERSUS WIRE LENGTHS

Rating of Equipment at 120 Volts (A)	Use Size 18 for Lengths (ft)	Use Size 16 for Lengths (ft)	Use Size 14 for Lengths (ft)	Use Size 12 for Lengths (ft)
5.8	1–125	126–150	151–190	191–295
7.2	1–50	51–150	151–180	181–230
9.8	1–50	51–125	126–150	151–180
13.0		1–50	51–125	126–165
15.0			1–50	51–150
20.0				1–50

Medium: Uses 16-gauge wire and has a current capability of 10 to 13 amperes. Typical appliances are a coffee maker, food mixer, electric knife, lamp, and fan.

Heavy: Uses 14-gauge wire and has a current capability of 13 to 15 amperes. Typical appliances are a room air conditioner, electric iron, television set, and dehumidifier.

Super: Uses 12-gauge wire and has a current capability of 15 to 20 amperes. Typical applications are a large air conditioner, portable electric heater, and large-capacity electric pot.

Cords are equipped with a plug at one end and a receptacle at the other. The three-connector type can be used with a two-terminal receptacle with the help of an adapter.

Precautions When Using an Extension Cord

An extension cord is a temporary device. Do not use it for permanent installations.

Do not use if either the plug or receptacle are damaged. Do not use if the cord is frayed, damaged, or worn.

Keep the extension cord away from water.

Do not change the cord in any way by cutting it, splicing it to add another length of cord, or modifying it in any way.

If any part of the cord becomes hot during use, replace the cord with one having a higher current rating.

If the cord is old and is beginning to show cracks in its insulating outer covering, replace it. Do not try to salvage it; do not keep it; do not offer it to anyone. Junk it.

An extension cord is not a toy. Store it out of the reach of children.

Buy only cords that carry the UL logo either on the package or on the product. Always buy a three-conductor type.

SINGLE WIRE COLOR CODING[21]

Single wire does not have its insulation color coded. Cables used for branch wiring are color coded as indicated in Table 3–10.

In a three-wire line the two hot leads are either both black or one may be black and the other red. The National Electrical Code, Section 210-5, requires that the (neutral) wire should be identified by a white or natural gray color. The Code re-

---

[21](*NEC: Article 800–2; page 852*)

TABLE 3–10.   WIRE COLOR CODING

Number of wires	Color coding
Two	White, black
Three	White, black, red
Four	White, black, red, blue
Five	White, black, red, blue, yellow

quires that a white wire must never be used for any purpose other than neutral. One requirement for this wire is that it must never be opened for any purpose, such as for a switch or fuse. A separate ground wire can be bare or color coded green.

DC RESISTANCE[22]

Whether a wire has an alternating current or a direct current flowing through it, the wire has a certain amount of opposition to the flow of either of these currents, which is referred to as its *dc resistance,* or simply its resistance.

The resistance of a wire is measured in terms of its opposition to the flow of a direct current through it and is generally measured per thousand feet at 68°F (20°C). The temperature is specified since the resistance of wire increases with temperature. In terms of gauge numbers, the resistance of copper wire becomes larger for each increase in gauge number.

Fusing Currents of Wires

An excessive flow of current through a wire can not only result in a fire but may also cause the conductor to melt. The amount of current flow that can result in melting of the conductor is known as the *fusing current* and appears in Table 3–11. Wires of other metals such as aluminum, German silver, iron, and tin are also susceptible to a fusing current, but are not included here since they are not involved in practical wiring.

CORDS[23]

The wire used in an electrical installation can consist of semirigid types such as Romex or BX, also known as in-wall types. Flexible wires, called cords, are used for connecting appliances and lamps to receptacles. There are three categories of

[22]*(NEC: 422–8; page 380)*
[23]*(NEC: Article 422–11; page 381)*

TABLE 3–11. FUSING CURRENTS OF COPPER WIRES IN AMPERES

AWG gauge	d (in.)	Copper $K = 10,244$
40	0.0031	1.77
38	0.0039	2.50
36	0.0050	3.62
34	0.0063	5.12
32	0.0079	7.19
30	0.0100	10.2
28	0.0126	14.4
26	0.0159	20.5
24	0.0201	29.2
22	0.0253	41.2
20	0.0319	58.4
19	0.0359	69.7
18	0.0403	82.9
17	0.0452	98.4
16	0.0508	117
15	0.0571	140
14	0.0641	166
13	0.0719	197
12	0.0808	235
11	0.0907	280
10	0.1019	333
9	0.1144	396
8	0.1285	472
7	0.1443	561
6	0.1620	668

flexible cords: lamp cord, heater cord, and heavy-duty cord (sometimes called power cord). An extension cord is a type of power cord. (*NEC: 422–8, page 380*)

Lamp and Power Cords

There are many types of lamp and power cords. The most common are the single paired, rubber insulated, and twisted pair. Twisted-pair cords consist of two cotton-covered conductors, which are coated with natural or synthetic rubber and then a final covering of cotton braid. Heater cords also belong to the flexible power cord family, except that the first coating around the copper conductors may be asbestos (more rarely used now) or some equivalent nonconducting, nonflammable material.

Heavy-duty or hard-service cords are usually supplied with two or more conductors surrounded by cotton and rubber insulation. When manufactured, the conducting wires are covered with an insulating material and then twisted, with the assembly covered by an outer layer of rubber.

Although referred to as lamp cord, this type is used for fans, lamps, radios, and TV sets. The wires are covered with a plastic insulating material. There are two solid conductors of No. 18 wire. Stranded wire is used more often, and this can consist of 41/34 wire, that is, 41 strands of No. 34 wire. The cord has a current rating of 7 amperes.

Various methods are used to help identify each lamp cord conductor. They may use different colors for the outside insulation. Some have a polarity ridge over one conductor. The cord is rip-type construction, meaning it is possible to separate the conductors by holding the two conductors and pulling them apart. The cord will then split down the center of the insulating material, facilitating connections with the individual wires. The identification letters for this type of lamp and power cord are SPT.

JOINING WIRES[24]

There are three ways of joining wires. The simplest connection is joining a pair of wires whose ends have been stripped of insulation. The wire ends can be twisted together, forming either a splice or a joint. This method can lead to problems if the connection becomes loose or the wires separate. Soldering the connection is more desirable, but is not always feasible if the connection is to be made in a hard-to-reach place or if the job is to be done in a hurry. Soldering requires some expertise and if not correctly done is worse than no soldering at all. In a worst-case situation, the resulting joint is cold soldered, with the heat of the iron causing the insulation to crawl away or become burned. For connections that use very small currents and low voltages, shrink wrap can be used. The wrap is heated after the connection is made and is used to cover the joint with a layer of insulation.

WIRE SPLICES[25]

A splice consists of joining wire conductors. To do so, it is first necessary to remove about 3 inches of insulation from the wire ends. A splice is both an electrical and a mechanical connection; it is electrical since it is necessary for the joint to have minimum resistance, and it is mechanical since the joint must be strong.

[24](NEC: Article 300–16; page 169)
[25](NEC: Article 300–22; page 171)

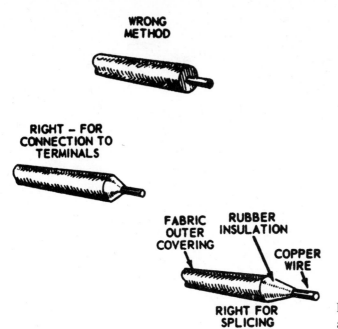

Figure 3–11. Method of stripping a wire for splicing.

Basic Splice[26]

Figure 3–11 shows how wires are prepared for splicing. The wires are cut so that the insulation tapers toward the central conductor and looks somewhat like a sharpened pencil. The wire must be shiny clean, with all corrosion removed. The wire should have no trace of any insulating material such as enamel. Scrape the wire with a penknife until the copper shines.

Figure 3–12a shows the next step in making the splice known as a rattail or pigtail. Twist the wires together, preferably using a pair of gas pliers and a vise. With the wires held tightly, turn the wires with the pliers.

It is much easier to work with solid conductors than with stranded wire.

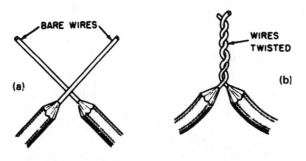

Figure 3–12. Basic splice known as a rattail or pigtail.

[26](*NEC: Article 305–2; page 173*)

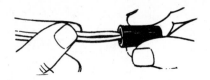

Figure 3–13. Bare wires protected with wire nut.

With stranded wire there is always the possibility that one of the strands will be removed accidentally. Count the strands to make sure they are all available and then twirl them to form the equivalent of a single conductor. If any strands are missing, cut the wire once again to make sure all the strands are present.

Clean the stranded wire with fine sandpaper or scrape it carefully with a knife or single-edge razor blade. The strands should have the same shiny appearance as solid conductor. Put the wires in a vise and twist the wires together as tightly as possible using a pair of gas pliers.

Joining the wires is a two-step process. After completing the first half of the twist, work on the second half. The final appearance of the joint should resemble that shown in Figure 3–12b.

Generally, when wires are connected they are covered by a wire nut, as shown in Figure 3–13. However, if this is not possible, cover the joint with plastic electrician's tape. Make a very tight progressive wrap, starting with the insulated end of the first wire; continue across the splice (Figure 3–14) and finish with a wrap around the insulated portion of the second wire. Wrap the tape at an angle so that each successive turn covers part of the preceding turn.

Older electrical tape, known as friction tape, was made of a fabriclike material coated with adhesive on one side. Its disadvantage was that it unraveled in time, and so the joint could lose some of its protection. In older installations it is advisable to replace friction tape with plastic electrician's tape.

The best thing to do about wire splices is to try to avoid them. When setting up a branch line from a service box to a receptacle, it is much better to use a continuous wire rather than one made up of splices, even though it may seem more economical to use scrap lengths. Do not pull a splice through conduit nor permit a splice to remain inside conduit.

To check a completed splice, hold each of the joined wires about 4 inches from the splice. Move the wires back and forth while watching the splice. Neither of the wires forming the splice should move, that is, they should not rub against

Figure 3–14. Technique for making an electrician's tape wrap.

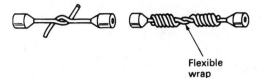

Figure 3–15. Splice with flexible connection.

Flexible wrap

each other, nor should it be possible to move them out of position. There should be no open spaces between the wires forming the splice.

There are various types of splices; with the selection a matter of experience and personal preference.

Flexible Splice

The problem with the basic splice is that the area of connection tends to be very stiff. This may or may not be a problem depending on the purpose of the splice.

The flexible splice is very much like the basic splice except that there is more space between the left and right connections. In this area the connecting wires are not twisted tightly together, and so there is more flexibility. Figure 3–15 illustrates the flexible splice.

Western Union Splice

As in the case of the basic splice, remove the insulation and clean the conductors. Bring the wires to a crossed position, as shown in Figure 3–16a. Take one of the wires and wrap it tightly around the straight portion of the other wire as in drawing b. Don't try making a finger wrap. Use a vise and a pair of gas pliers and make the wrap as tight as possible. Now take the second wire and wrap it around

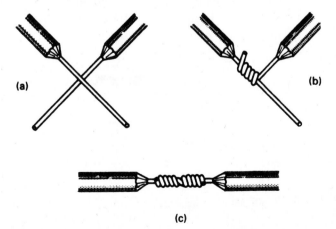

(a)

(b)

(c)

Figure 3–16. Western Union splice.

Figure 3–17. Staggered splice.

the first. The result will be a joint as in drawing c. Test the connection by pulling on the joined wires to detect any possible looseness.

Protect them with a wrap of plastic electrician's tape.

Staggered Splice

It may be necessary to splice a pair of adjoining wires with the possibility of a short between the joints. Still another problem is a joint that is large and bumpy. This can be avoided by using a staggered splice, offsetting the splices as indicated in Figure 3–17.

After completing the staggered splices, tape each individually and then put tape over both. Tape tightly with plastic tape. Don't let the tape become slack when using it.

Since splices do not permit the use of wire nuts, it may be desirable to solder each joint. It is easier to complete one splice, solder it, and then move on to the next splice. Separate the two wires when soldering to avoid burning the insulation. Avoid dropping hot solder on the insulation.

Stranded Wire Splice

Splicing stranded wire requires more care and patience than in splicing solid conductors. A method for making such a splice is shown in Figure 3–18.

Separate the individual strands and fan them outward. The wires are very thin and are easy to break. Bring the two wire ends together so they mesh, as in drawing a. Make sure they intertwine as much as possible. Bring the wires back as in drawing b. Twist the wires around each other as tightly as possible as in drawing c. Make sure no individual strands stand away from the others.

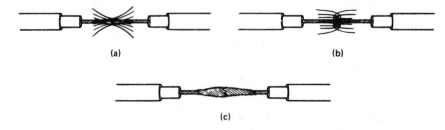

(a)   (b)

(c)

Figure 3–18. Stranded wire splice.

JOINTS

The purpose of a splice is to join wires so they form a continuous straight line. A joint is also the joining of wires, but the spliced section forms an angle with the original wires that may be as much as 90°. However, the two terms, joints and splices, are often interchangeable.

Fixture Joint

A fixture joint is so called since it is commonly used for connecting light fixtures to the branch circuit of an electrical system. Generally, in such installations, the wire supplied with the fixture is smaller in diameter; that is, it has a larger gauge number than the branch wire, the wire that will supply electrical current to the fixture.

Neither the rattail joint nor the fixture joint can withstand much mechanical strain. It is essential to remember that it is not the function of the wires to support the fixture. There should not be any stress or pull on the joint after it is completed.

The steps for making this joint are illustrated in Figure 3–19. First, as in drawing a, strip and clean the wires and then cross them. The wire at the left is the fixture wire; that at the right is the branch wire. Wrap the fixture wire tightly around the branch wire as in drawing b.

Take the branch wire, as in drawing c, and bend it over the fixture wire. This will form a U-shaped hook. Finally, as in drawing d, take the fixture wire and wrap it around the hook. The joined wires can now be covered with a wire nut.

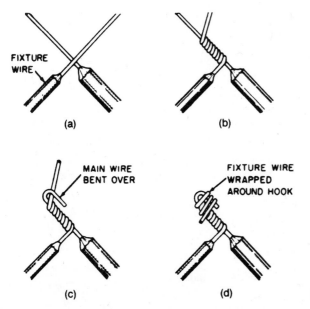

Figure 3–19. Fixture joint.

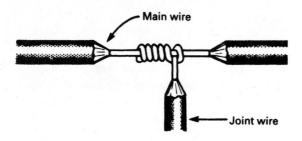

Figure 3–20.   Knotted tap joint.

Wire nuts come with various-size openings, and the nut may need to have a large opening because of the bulk of the joint.

Knotted Tap Joint

This joint is illustrated in Figure 3–20. To make it, strip about 1 inch of insulation from the wire identified as the main wire, to which the joint wire is to be connected. Strip about 3 inches of insulation from the joint wire. The two wires will look like the capital letter T, when completed, with the main wire horizontal and the joint wire vertical.

To make the knotted tap joint, cross the two wires at right angles. Bend the joint wire over the main wire and then under the main wire to form a knot. After making the knot, continue wrapping the branch wire around the main wire in tight turns.

It is possible to eliminate the knot. It all depends on whether the wires will be subject to any kind of strain or slip. If there is no mechanical strain, forget about the knot. The knot, however, does make the tap much more mechanically secure.

Stranded Wire Tap

Like solid conductor, stranded wire can be used to form a tap, as shown in Figure 3–21. As a first step, select an area along the wire to which the tap is to be made and separate the individual wires. Then strip the end of the wire that is to form

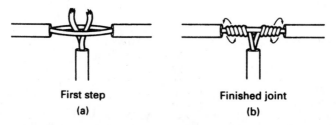

Figure 3–21.   Stranded wire tap.

part of the joint and insert it into the open space of the horizontally positioned wire, as indicated in drawing a. Separate the strands into two approximately equal groups and twirl them to make them as similar to solid conductor as possible. Wrap each group around the horizontal wire to form the tap as shown in drawing b. The finished joint can then be soldered and wrapped in plastic tape.

Three-wire Joint

Three wires can be jointed if the usual preparatory precautions are followed. The wires must be stripped and cleaned, and then twirled around each other so tightly that no individual wire can be moved. The connection must be covered with a wire nut so that it is tight on the joint and covers it completely. The nut can then be covered with plastic electrician's tape.

Joining Wires of Different Gauges

If two or three wires are to be joined in series, the maximum ampacity is that of the wire having the largest gauge number, that is, the smallest cross-sectional diameter. Any fuse or circuit breaker in series with the joint must have a maximum current rating based on the smaller current. If the wires have current ratings of 8, 10, and 15 amperes, then the fuse or circuit breaker should be an 8-ampere type.

WIRE CUTTERS AND STRIPPERS

A number of wire cutters and strippers are available, with some that are combinations. Diagonal cutters or knives can be used for cutting wires, but they are less than desirable for wire stripping and are capable of nicking copper conductors; in the case of stranded wire, they can easily cut through one or more.

Figure 3–22 shows a combination wire cutter and stripper. It can cut and strip clean all sizes of solid or stranded copper wire. This does not mean that a tool of this kind will automatically supply perfect results. If not used correctly, it can damage or cut through wires.

Figure 3–22. Combination wire cutter and stripper.

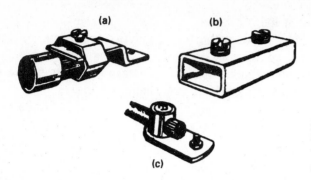

Figure 3–23. (a) and (b) Screw-type connectors; (c) Spring-lock connector.

WIRE CONNECTORS[27]

Wires that are intended to carry substantial currents may be too thick to splice and, even if joined, cannot be capped by a wire nut. Such wires can be joined through the use of solderless connectors, also known as wire connectors or wire caps. There are a number of types, and a few are shown in Figure 3–23. However, the preparatory work of stripping the wires and inserting and fastening them with the connector are similar.

To use the fastener, strip the wires and insert the bare ends using one screw as in drawing a or two screws as in drawing b. Usually, such fasteners are suitable only for solid conductors. The screws used in these two examples may be setscrews or regular machine screws. The machine screws are often flat-head types with a single slot.

The item in drawing c is a spring-loaded connector. Press down on the top of the connector, insert the wire, and then let go. A spring on the bottom of the connector will push up against the wire. The advantage of such a device is that connections can be made extremely quickly. However, it does present two problems. It is unsatisfactory for moderate or large currents, that is, for currents of half an ampere or more. Another disadvantage is that in time the spring will become weaker, reducing the tension and making the connection poorer.

[27](NEC: Article 517–34; page 581)

Chapter 4

Electrical Components

PART I

Service Box, Fuses, Circuit Breakers, GFCI, Plugs, and Receptacles

Electrical power is delivered from a service box to a number of receptacles through various branches; this is accomplished with the help of a number of accessories to assure that the current reaches its proper destination. These accessories include fuses, circuit breakers, GFCIs (a special type of circuit breaker), receptacles, electric boxes, plugs, switches, single and multiple taps, receptacle strips, screws, straps, and staples. Some are involved directly in the current path, while others supply both control and support.

The number and variety of these electrical components and a listing occupy many catalog pages. The description of components in this chapter cannot possibly cover them all, but it can lead to an awareness of what is available.

SERVICE BOX[1]

The service box, also known as the service panel, distribution box, fuse box, or circuit-breaker box, is the area of entry of electric power into a home, business, or farm by means of power lines supplied by an electric utility (Figure 4–1). Prior to the point of entry, including the meter, the lines are the responsibility of the utility. In a private home the box is usually located in the basement; in an apartment it is in the kitchen (Figure 4–2).

The service box has several functions. In this box the different branch circuits (Figure 4–3a and b) supplying electric power to the various receptacles have their beginning (*NEC: Article 100; page 24*). Branch circuits are used to distribute the load so that the current carried by the wires of the branches does not exceed their ampacity. The input cables to the service box make a tremendous amount of power available. The purpose of the service box is to divide that power into smaller segments, and these are handled by a number of branches. The amount of electrical power distributed by a branch depends on the load, that is, the number of appliances serviced by that branch, the amount of operating power that they require, and whether they are turned on or off.

Branches (*NEC: Article 210–1; page 52*), are used as electrical networks so that if one is disabled the others are not affected. This arrangement makes servicing easier since the electrical problem is confined to a limited area.

By using fuses or circuit breakers, each branch is individually protected, permitting each to carry a lesser or greater current. Some branches can use a smaller or larger gauge of wire, depending on the load.

There are two types of switches in service. These include the switches for the individual branches exiting from the service box. On the service box panel you will find numbers corresponding to the branches. The switches are marked with numbers corresponding to the maximum amount of current per branch.

The main switch in the box can be used to stop the flow of current to all the branches at one time. This does not mean that all input voltage to the box is eliminated. The input from a utility's service lines always remains connected.

Branch Circuit Wiring[2]

Branch circuits differ in several respects: in the current ratings of their protective fuses or circuit breakers, in the gauge of the wire they use, and in the number of receptacles they supply. Small TV sets, electric clocks, electric shavers, small

[1](*NEC®: Article 100; page 33*)
[2](*NEC: Article 100; page 25*)

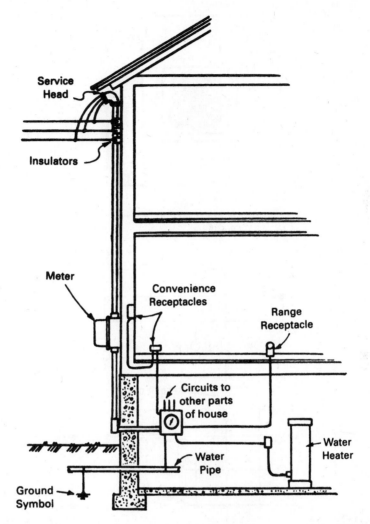

Figure 4–1. Wiring on the outside of a building consists of three wires color coded red, black, and white. These are fished through galvanized rigid conduit and are brought out through a service head at the top of the conduit. The three wires are then spliced onto the utility company's power line. (This illustration and accompanying text are also supplied on page 8.)

electric fans, electric pencil sharpeners, and other limited-current appliances operate from the 120-volt line, use a fuse or circuit breaker rated at 15 amperes, and use No. 14 gauge wire between the service box and the receptacles. The maximum total power is obtained by multiplying the line voltage by the maximum current. Thus, $120 \times 15 = 1800$ watts.

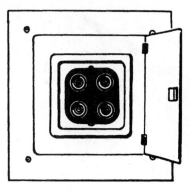

Figure 4–2. Fuse cabinet of type used in apartments. This box does not have a main fuse or switch. The fuses are only for branches extending into the apartment. More modern boxes use circuit breakers.

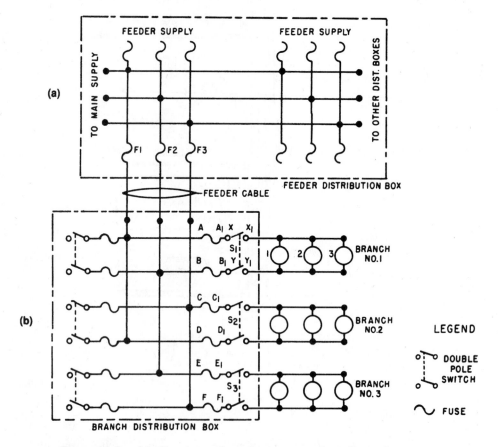

Figure 4–3. (a) Circuits or branches connected to the main power lines. The hot leads of circuits 1 and 3 are connected to the red wire; other circuits to the black wire. All the branches are wired to receptacles. (b) Wiring of connections to a service box using fuses. The wiring is similar for a service box using circuit breakers.

Another branch, protected by a 20-ampere fuse or circuit breaker, uses No. 12 gauge wire. The maximum power-handling capacity is $20 \times 120 = 2400$ watts. The difference here is simply a greater power-handling ability. The source of electrical energy has not changed and is supplied by the feeder lines from the power utility. This branch is intended for electric irons, portable electric heaters, and electric toasters. Note that these are all heating devices.

Advantage of 240 Volts

To obtain a power figure, multiply voltage (in volts) times current (in amperes). To maintain the same amount of power but also reducing the current, double the voltage and use only half the current.

$$120 \text{ volts} \times 10 \text{ amperes} = 1200 \text{ watts}$$
$$240 \text{ volts} \times 5 \text{ amperes} = 1200 \text{ watts}$$

Consequently, a smaller-gauge wire can be used for current-hungry appliances, such as electric ovens, washers, and dryers, by operating them from a 240-volt branch line.

The voltage supplied to the service box is 240 volts, 60 Hz. This voltage consists of 120 volts on either side of a neutral wire, making it possible to use either 120 or 240 volts, so a branch line can be either of these voltages. Voltage is always measured between two points, one of which is known as the *reference*. For the service box, the neutral or white wire is the reference or zero voltage point. A black wire is 120 volts with respect to the white wire. There is also a red wire and it, too, is 120 volts with reference to the white wire. The voltage between the red and black wires is 240 volts.

Master Fuse or Circuit Breaker[3]

The two hot leads, black and red, coming into the service box are wired to a pair of main fuses or circuit breakers, followed by a main or master disconnect switch (Figure 4–4). This switch is capable of turning off all power supplied to the various branch circuits. In older homes the disconnect switch is outside the service box; subsequently, the switch was incorporated inside the box.

Low voltage (120 volts) is used for lights and relatively low power appliances. The higher voltage (240 volts) is for dryers, electric stoves, and washers.

Each fuse or circuit breaker is numbered and also has data about its current-carrying capacity. On the inside panel of the service box is a listing, indicating the fuse or circuit-breaker number and the room or rooms, areas, or appliances it controls.

[3](*NEC: Article 240–80; page 118*)

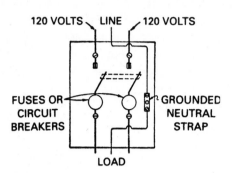

120 VOLTS   LINE   120 VOLTS

FUSES OR → CIRCUIT BREAKERS

GROUNDED NEUTRAL STRAP

LOAD

Figure 4–4. Switching arrangement for input to service box. Note that the neutral wire is not fused. Each hot wire is 120 volts with respect to neutral. The two main fuses (or circuit breakers) must carry the total current demand and are rated from 30 amperes to as much as 200 amperes. These rarely need replacement (fuses) or adjustment (circuit breakers).

FUSES[4]

A fuse (also called an overcurrent device) has two purposes: to protect a branch line, the line leading from the fuse box to various receptacles, and to protect electrical appliances plugged into those receptacles. A short circuit can occur when the hot lead of a branch line touches either the ground lead or the neutral wire of that line.

When a fuse *blows,* the term used for the opening of a fuse, the effect is similar to that produced by an open switch. All current in that branch line stops, and any appliances plugged into receptacles in that branch line stop functioning: lights turn off, radio and television receivers stop working. In most cases, other branch lines are not affected.

Plug Fuse[5]

The plug fuse is one of the more common types and is so called because its base resembles that of an electric light bulb, as shown in Figure 4–5. It contains a small, metallic ribbonlike alloy of lead and tin designed to melt at a specific

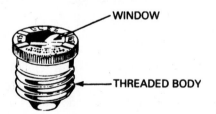

WINDOW

THREADED BODY

Figure 4–5. Plug fuse.

[4](*NEC: Article 240–40; page 116*)
[5](*NEC: Article 240–50; page 116*)

amount of current. The ribbon melts from the heat that is generated and, since it could possibly spatter, is enclosed in a strong housing.

Plug Fuse Ratings[6]

Fuses are rated by their current-carrying ability, commonly from about 5 amperes to 30, although some are rated for less than 5 amperes, and types other than the plug fuse may be rated at more than 30 amperes. A common home type of plug fuse is the 15-ampere fuse used for protecting No. 14 gauge wire. As long as the current flow through that wire remains at 15 amperes or less, the fuse will remain as a good connecting link. All fuses are connected in series with their respective branch lines, never in parallel (shunt). A fuse put in parallel with a branch line will blow immediately.

Since fuses are rated by their current-carrying ability, it gives the impression that it is the current that is the controlling element. The current rating is simply a convenient method of identifying fuses. A fuse will open because it exceeds its power-handling capability.

Consider a 10-ampere fuse whose resistance is 1 ohm. Its power-handling ability is based on the formula

$$P = I^2 R$$

and this is the same as

$$P = I \times I \times R$$

where P is the power in watts, I is the current in amperes, and R is the resistance in ohms. Thus, the fuse's power-handling ability is

$$P = 10 \times 10 \times 1$$
$$= 100 \text{ watts}$$

If the current flow is increased to 40 amperes, perhaps by plugging too many active appliances into receptacles in its branch line, then the power at the fuse will be increased. P now becomes

$$P = 40 \times 40 \times 1 = 1600 \text{ watts}$$

The power supplied to the fuse is now 16 times the original amount. The only way the fuse can dissipate this power is by converting it to heat, and so the fuse element literally catches fire.

[6](NEC: Article 240–50 (b); page 116)

Types of Fuses[7]

There are a large number of different types of fuses with a current protection capability ranging from a fraction of an ampere to as much as 6000 amperes. Operating voltages can be 2.4, 4.8, 5.5, 7.2, 12, 24, and 150 with some ranging as high as 36 kilovolts. Fuses are used for dc, ac, and universal motors.

The following is a partial listing of various types of fuses; new applications are constantly being developed. In some instances the fuses are not commercially available but are tailored for a specific application. This listing does not include circuit breakers.

Time delay
Fast acting
Multipurpose
One time
Renewable
Medium voltage
Cartridge
Slow blow
Plug
Current limiting
Round semiconductor
Square semiconductor
General purpose
Knife blade
Snap in
Fuse to circuit breaker adapter

Fuse Location[8]

For electric appliances, the fuse is located in the service box. However, for electronic equipment such as TV sets and high-fidelity equipment, the fuse can be built into the units. The fuse is generally contained in a small, plastic fuse holder mounted on the rear apron and is readily accessible for replacement (Figure 4–6). The fuse is wired in series with one side of a power cord that is connected to the ac line.

[7]*(NEC: Article 240–51; page 117)*
[8]*(NEC: Article 240–30; page 115)*

Figure 4–6. Type of fuse used for electronic equipment.

Time Delay

The flow of current in a branch line is variable, not only depending on the number of appliances being used, but also on the varying current demands of those appliances. When electric lights are first turned on, their resistance is low and their demand for current is high. As they become hot, their resistance increases and so their current usage decreases. This action lasts a very short time, measured in seconds.

For this reason, fuses have a time lag; that is, they do not blow instantaneously. Thus, a 15-ampere fuse could carry a current of possibly 20 amperes, provided this higher current remained for only a very short time. This could be a momentary overload. An uninterrupted flow of 20 amperes would heat the fuse in a short time and it would open. For much higher currents, such as 50 or 100 amperes, the fuse would open practically at once. A good rule to follow is that the fuse rating should never exceed the current-carrying capacity of the branch wire that it protects.

Basically, a fuse contains a short length of a conducting material designed to melt when a certain amount of current flows through it. Since the melting process can cause the conductor to splatter, it is enclosed in a strong housing, such as metal, plastic, or fiber.

The plug fuse has a window at the top covered with a transparent material through which it is possible to see the fuse element. A black or discolored smear is evidence of an open fuse. These fuses are sometimes available in two different shapes, hexagonal and round. If the fuse is 15 amperes or less, the window has a hexagonal shape, while for fuses of more than 15 amperes it is round.

Replacing Plug Fuses[9]

All fuses, including plug types, have an amps rating. Their current ratings are 3, 6, 10, 12, 15, 20, and 30 amperes. When replacing a fuse, always use one having the same ampere designation.

Identifying Fuse Ratings[10]

The rating of plug fuses is often stamped in the form of a number on the fuse base, or it may be printed on a card fitting against the inside of the fuse window. If the rating is unknown it is always better to discard the fuse.

[9](*NEC: Article 240–24; page 115*)
[10](*NEC: Article 240–50; page 116*)

Overload Tolerance

Whether or not a plug fuse will open depends on the amount of excess current, the time duration of this current, and the speed with which heat can escape. For a short-circuit condition, a fuse will open practically immediately. With an overload condition, the fuse may or may not blow, depending on how long the overload lasts. As a rule of thumb, if the overload is 50%, the fuse will open in 1 to 15 minutes. Thus, with a 10-ampere fuse a current of 15 amperes will cause it to open, not at once, but within the 15-minute period, depending in part on the ambient temperature. In winter, if the fuse box is extremely cold, an open-fuse condition will take place closer to 15 minutes. In very hot weather, an open will take less time.

Fuse Deterioration

Fuses do not deteriorate in storage, so it is well to have a supply on hand consisting of at least one each of the different plug fuses used in the service box. If the box is in a dark location, it will be helpful to have a battery-operated flashlight nearby if normal natural and artificial light is inadequate.

Identifying Fuse Types

Fuses can be identified in various ways. One of the most common is by their current rating. Some fuses are time-delay types, while some, such as the S fuse, use a letter. And some fuses are recognized by a trade name or by their design, such as the circuit-breaker fuse, or by a special shape, such as the cartridge fuse.

Time-delay Fuse

In some circuits a momentary overload current is a normal condition. A motor, for example, can be practically a short circuit across a branch line when the motor is started. As the motor armature begins to turn, it develops a counter or opposing voltage known as a counter electromotive force or counter emf. This counter emf opposes the line voltage and in so doing reduces the current taken by the motor to some safe value. The starting current of a motor might be 30 or more amperes, but in operation it will require only 5 amperes. If an ordinary fuse is used, it may blow under these conditions. For circuits using motors, the fuse (or fuses) are often time-delay types.

Figure 4–7. S-fuse.

S Fuse

The S fuse (Figure 4–7) is a two-part type. The first part consists of a base that re-sembles that of an electric light, and this is the part that screws into a socket in the fuse box. This first part also has an opening to permit the insertion of the fuse holder. This second part, an adapter, screws into the opening. It has a narrow di-ameter and for that reason cannot be inserted incorrectly into any other fuse-holder socket.

The type S fuse is sometimes called nontamperable. Its purpose is to prevent the use of fuses whose current rating exceeds the ampacity of the gauge of wire used in the branch line that it is supposed to protect. The problem with ordinary plug fuses is that all their bases are interchangeable; that is, a 30-ampere plug fuse will fit into the socket normally occupied by a 10-ampere fuse.

The adapter used by a type S fuse is designed to accept only fuses having the correct current rating. Thus, a 15-ampere S fuse will only accept an adapter for 15-amperes. All type S fuses are also time-delay types. S fuses are different in another respect. The fuse must be inserted into its adapter strongly so that it makes very firm contact; otherwise, the fuse will act like an open circuit and not permit current flow in its branch circuit. While modern home construction now uses circuit breakers, if plug fuses are used, the NEC requires the S type.

Fuse Pullers[11]

Plug fuses can be removed by gripping and turning the insulated portion sur-rounding the window. There is the possibility of shock if the fingers slip and touch the ferrule. This can be avoided in two ways. One method is to shut off the power at the main switch, but this will cut off power to all appliances, including clocks, necessitating resetting them. A better technique is to use a fuse puller (Figure 4–8) or else to cover the fuse with a dry cloth.

Service boxes are often located in basements where moisture is commonly present. A good safety procedure is to stand on some dry insulating material, such as a section of rug. Work gloves, if dry and clean, are also helpful. Using both the rug and gloves is preferable.

[11]*(NEC: Article 240–54; page 117)*

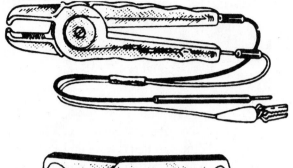

Figure 4–8. Cartridge fuse pullers. The puller at the top is also equipped with a neon glow lamp for testing, plus a pair of test leads.

Cartridge Fuses

The cartridge fuse (Figure 4–9) is available in two forms, renewable and nonrenewable. The nonrenewable type consists of a cylinder made of a hard, fiberlike material containing the fuse element. It melts at a predetermined current value and is either soldered to or mechanically fastened to a pair of metal ferrules, one at each end of the fuse housing. The ferrules are end caps and are used for inserting the cartridge fuse into a pair of spring metal holders.

Cartridge fuses are generally designed to handle larger currents than plug fuses, but they are available in small current values as well. They have current ratings of 3, 6, 10, 25, 30, 35, 40, 50, and 60 amperes. Up to 30 amperes the fuses are 2 inches long, and for currents between 30 and 60 amperes they are 3 inches long.

For a renewable cartridge fuse, the fusible metal strip can be replaced after it has opened. To insert a new fusing strip, remove the ferrules, mount the strip, and then replace the ferrules.

Knife-blade Fuses

Knife-blade-type cartridge fuses have much higher current ratings than the ferrule type. These can have current ratings of 60 to 600 amperes. Knife-blade cartridge fuses from 60 to 100 amperes are $7\frac{1}{8}$ inches long; those having higher current ratings are longer.

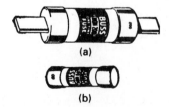

(a)

(b)

Figure 4–9. (a) Knife-blade cartridge fuse and (b) snap-in fuse.

ELECTRICAL FAULT LOCATION

The service box can be used as an aid in servicing. The starting point in locating an electrical problem is a matter of personal preference. Start either at a suspected receptacle or at the electrical box.

If an appliance stops working, try these steps:

1. Make sure the plug of the appliance is in its receptacle and that it fits firmly. Move the plug and turn it slightly left or right with the appliance switch on. Sometimes the plug is removed accidentally. Try the appliance in a different receptacle. If the appliance is too heavy or too difficult to move, use an extension cord. Insert a test lamp into the receptacle to see if it is receiving power. The test lamp can be any convenient, easily portable household lamp or a more convenient neon lamp continuity tester (Figure 4–10).

2. If there is no power at the receptacle and if the tester does not work, try it in other receptacles in the same room and adjoining rooms. If other receptacles do not supply power, then the fault can be at the service box or somewhere in the line between the service box and the open receptacle or between a pair of receptacles.

3. Check the fuses for evidence of blowing. A fuse whose window is clear but whose fuse strip is melted is evidence of an overload condition, but the overload is a light one. If a plug has a badly smudged window, the overload is severe. This could also be an indication of a short circuit, a maximum overload condition.

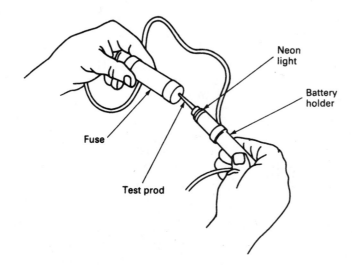

Figure 4–10.   Continuity tester.

4. Try to locate the possible cause of the overload. Remove all appliances connected to the suspected branch line. Replace the fuse. Test the line by inserting the tester in the receptacle that originally held the plug of the inoperative appliance. The lamp should light. If not, check the replacement fuse. If it has blown, there is a short in the line between the service box and the receptacle.

5. If the test lamp lights, plug in each appliance. Check the fuse each time. The appliance that causes the fuse to blow once again has some defect.

6. If none of the appliances causes the replaced fuse to open, and all the appliances work, there are three possibilities:
   a. One of the appliances caused a momentary overload.
   b. One of the appliances is motor operated. Replace the fuse with a time-delay type.
   c. The load on the line is at maximum or a little above. The cure is to connect one of the appliances to some other line, that is, to redistribute the load, or to rewire so that the line can carry a stronger current.

7. If the tests indicate a possible short in the branch line wire, trace the wiring to determine if there are any splices. There may be a short in one of the receptacles connected to the branch line. If the appliance works intermittently, make sure the plug fuse is seated firmly. If a cartridge fuse is used, make sure the end ferrules are firmly gripped by its metal holders. Check the cartridge fuse by gripping the fiber cylinder portion and try to move it. If it moves easily, take the cartridge out and tighten the holders. Make sure the power is off before doing this.

CIRCUIT BREAKERS[12]

Fuses are found in older buildings. More recent homes use circuit breakers (Figure 4–11), a current-sensitive switch. When a circuit breaker opens, equivalent to a blown fuse, all that is needed is to correct the fault and to reset the breaker, which is done by actuating the switch handle of the breaker, (*NEC: Article 240–80; page 118*).

Circuit breakers are also available with a time-delay feature and so can be used with motors having a high starting current and a lower operating current.

Operational Safety

Circuit breakers, and fuses in much older service installations, have a double function. They warn against short circuits and overloads. They can do this properly provided the accompanying branch wiring and loads are selected correctly.

[12](*NEC: Article 240; page 106*)

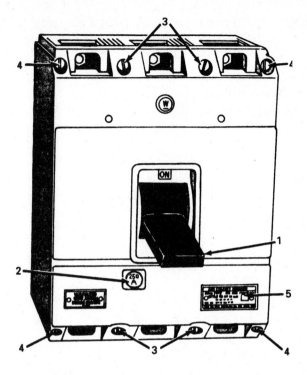

Figure 4–11. Circuit breaker showing essential parts: (1) operating handle, (2) ampere rating, (3) mounting screws, (4) cover screws, (5) breaker nameplate.

(1) operating handle;
(2) ampere rating;
(3) mounting screws;

(4) cover screws;
(5) Breaker name plate.

Thus, if a load has a 15-ampere rating and is connected to a branch wired with No. 12 gauge conductors rated at 20-amperes, the fusing of the circuit should be kept to 15 amperes. The component in a branch line having the lowest amperage rating should decide the current rating of the circuit breaker.

The current ratings of circuit breakers and branch wiring are fixed. Consequently, it may be easier to change the distribution of loads to keep branch wiring and circuit breakers within their current limits.

Not all loads are used at the same time. The concept of trying to remember this and to disengage loads is a poor practice.

Types of Circuit Breakers[13]

Like fuses, circuit breakers are constructed in various ways. They may be thermal or magnetic or work by a combination of thermal and magnetic properties.

[13](*NEC: Article 240–4; page 109*)

Thermal circuit breaker.   The thermal breaker has a bimetallic element made by bonding two different metals to each other. Each metal has a different temperature coefficient of expansion. This means the two metals are affected differently by increases in temperature, with one expanding more rapidly than the other. Since the metals are joined, a temperature rise will make the metals bend. The bimetallic element will then act as a latch, tripping the circuit breaker and opening it. The bimetallic element will do this when the current exceeds a predetermined value.

Once the breaker is open, current stops flowing since the bimetallic element works as a switch. The switch can now be reset to its closed position. If the condition that caused the excessive current flow is removed, the circuit breaker will remain closed. If the excessive current condition remains, the breaker will open again.

Magnetic circuit breaker.   This breaker works by energizing a coil, thus making it into a magnet capable of attracting a bit of ferrous metal that is part of a switch. As long as current through the magnetic circuit breaker coil remains normal, the magnet is not strong enough to pull the metal strip away from its closed-switch position.

In case of an overload or a short circuit, the current through the magnetic circuit-breaker coil increases substantially. The coil becomes a stronger magnet and attracts the metal strip. This strip is part of a switch and in moving toward the electromagnet opens a circuit.

Ordinarily, the metal strip would be pulled down immediately, closing the circuit once again, for two reasons. The first is that the metal strip is spring loaded and is connected to the switch by a spring. The spring tends to keep the switch closed. The second is that once the current is interrupted there is no longer a magnetic field around the coil, and so there is no way for the coil to hold the switch strip.

However, the magnetic circuit breaker has a tiny latching mechanism. When an excess current flows through the circuit breaker coil, it pulls the metal arm of the switch up toward itself. A small mechanical latch then moves into position to hold the metal arm against the pull of the spring.

Resetting the circuit breaker releases the latch and permits the spring to pull the metal arm toward itself, closing the switch and allowing current to flow in the circuit once again. Of course, if the overload condition is not removed, the circuit breaker will trip open again.

Magnetic circuit breakers are used in applications in which fairly heavy currents flow and are designed for stronger circuit conditions than the thermal type. The thermal breaker is more sensitive and more responsive to smaller currents.

Thermal-magnetic circuit breaker.   The thermal-magnetic breaker is a combination of thermal and magnetic principles and is used for installations in which there is a wide range of current flow. The thermal element protects against current surges and overloads in the lower-current range, while the magnetic element supplies protection against high-current shorts.

Circuit-breaker Converter

The circuit-breaker converter (Figure 4–12) converts fuses into circuit breakers. Installation is simple. Remove the fuse and screw its replacement converter into the fuse socket. The unit is capable of handling momentary overloads. The converter indicates an overload condition when its colored center button pops out.

If the converter opens, it can be reset following removal of the overload condition by pushing the centrally located button in. It is important to use a converter having the same current rating as the fuse it replaces.

Operating the Circuit Breaker

The circuit breaker looks like a switch, but it is partially semiautomatic, for it can turn itself off. When the current does not exceed its rated amount, the switch remains in position, but when the circuit breaker is overloaded it turns to its off position. However, the breaker must be reset manually.

For 120-volt branches, the breaker has a single on-off toggle. For a 240-volt branch, it consists of two breakers having a joined handle. Both sides of this handle open together and must be operated as a unit to close. Typically, a circuit breaker can be reset by pushing the handle to the off position, as far as it will go, and then moving back to on.

Not all breakers are reset in the same way, but a little experimentation will soon reveal the technique. Some breakers indicate a tripped position when the toggle moves to its center. Others show a red marker when this happens. In either case, move the toggle to its full off position and then move it to on. The current rating of the toggle is shown by a number engraved or printed on the head of the toggle.

Figure 4–12.   Plug-in circuit breaker replacement for a fuse.

Advantages of Circuit Breakers

Circuit breakers have advantages over plug fuses, so much so that they are not only used in new construction, but service panels using breakers are selected when wiring is updated.

There is no way the user can tamper with the thermal or magnetic element of breakers since they are sealed. This means that, unlike fuses, there is no way of altering the current range. Unless the breaker becomes defective, there is no need for replacement if the breaker opens, nor is it necessary to keep a supply on hand. There is also no possibility of shock from handling since the voltage connections are behind a panel.

Home-type circuit breakers have a main breaker used for disconnecting power to an entire home. The main breaker is followed by a series of breakers, one for each 120-volt branch line and a pair for 240-volt lines. A typical service box will have between 10 and 20 breakers.

When the Breaker Opens

For trouble with an appliance, its receptacle, or its branch wiring, follow the same procedure described for fuses. Make sure the breaker is in its on position. If it is, then the problem is either in the appliance, one or more receptacles, or the branch line.

GROUND-FAULT CIRCUIT INTERRUPTER[14]

The ground-fault circuit interrupter (GFCI) is a member of the circuit-breaker family. The current that flows through the hot lead (the black lead) of a wiring system also flows through the white (neutral) lead. However, if through some fault in this system more current passes through the neutral line, this current will try to find its way to ground.

The neutral and the ground leads are parallel conductors, with the ground lead connected to earth. If both conductors have equal amounts of resistance, the same current will pass through them. The neutral (white wire) is an uninterrupted, continuous conductor, but the ground lead may be cut and connected at various points, adding some resistance. Touching the ground lead when an unbalanced current condition exists means that an additional path has been provided for the current, thus presenting a dangerous condition, referred to as a *ground fault*. The current imbalance is very small, and even if the current flow through the black and white leads is on the order of amperes, the imbalance is measured in mil-

[14]*(NEC: Article 210–8; page 57; Article 215–9; page 68)*

liamperes. However, to the human body this is an intolerable amount, and touching a presumably safe ground can result in electrocution.

Ground faults occur through current leakage to ground from electrical appliances and most often occur where moisture is present, as in the bathroom, the kitchen, basement, workshop, and laundry room. Outdoors, ground faults occur in swimming pools, electric barbecue grilles, motorized devices, bug zappers, exterior lighting, Christmas tree lights, and power tools, (*NEC: Article 210–8; page 57*).

A GFCI is intended to stop the flow of leakage current and in its operation is practically instantaneous. The unit works by making a comparison between the current flow through the hot and neutral leads, and it does so continuously. If there is a difference, ranging from a few milliamperes, but possibly no more than 200 milliamperes, the GFCI will open the circuit and it will do so in $\frac{1}{40}$ second, or possibly even faster.

GFCI Types

One characteristic of electrical components is that there is usually a variety, and the GFCI is no exception. The simplest is the plug-in GFCI, but it supplies the least amount of protection. All that is required is to insert the GFCI into a receptacle and the electrical device into the GFCI. There are two disadvantages. It can only be used indoors, and it protects only against appliances plugged into it. Figure 4–13 shows the front of the plug-in GFCI.

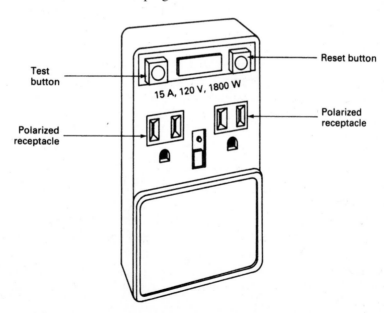

Figure 4–13.   Plug-in GFCI.

A different approach is to use the GFCI to replace an existing receptacle. It will then act to protect the entire branch line of which the receptacle is a part. However, the extent of its protection depends on where in the branch line the GFCI is inserted.

The greatest protection can be had when the GFCI is installed in the service box, but even here it will protect only the electrical branch in which it is connected. Ideally, a GFCI should be installed in each branch starting at the service box. The GFCI not only supplies electrical protection but also works as a circuit breaker.

The front view of the GFCI in Figure 4–13 shows that it has a provision for the use of two polarized plugs that have two blades plus a ground prong. There are also a pair of push buttons, one of which is used for testing the GFCI and the other for resetting the circuit breaker.

Line Cord GFCI

A power line cord equipped with a GFCI is available for outdoor use. The cord has a length of 6 to 25 feet and will trip the breaker off in a current range of 4 to 6 milliamperes. The GFCI is waterproof, while the power cord of which it is a part is covered with an insulation material that resists not only water but also sunlight and chemicals. This arrangement permits its use in boat yards and marinas, or with recreational vehicles. It also supplies leakage current protection when using power drills, table saws, floor polishers, battery chargers, wet/dry vacuum cleaners, and high-pressure washers.

The GFCI and its accompanying extension cord for outdoor use are waterproof and both types, direct plug in and extension cord equipped, are UL listed under Standard 943 for portable applications. A GFCI does not prevent shock. Its function is to limit its duration to a time period considered safe for healthy people. If a ground fault occurs, the GFCI device quickly interrupts the current before it can harm the user of the plugged in appliance.

A GFCI can be a plug-in device or it can be had as an integrated unit, combined with a receptacle. In the latter case, separate specifications are supplied for the GFCI and the receptacle. A typical GFCI could be rated at 15 amperes 120 volts ac, while its accompanying receptacle could be 20 amperes, 120 volts ac. For the GFCI the trip level would be 4 to 6 milliamperes.

Like the separate GFCI, the integrated unit offers a feed-through capability providing protection for receptacles located downstream. The construction of the receptacle is for a standard $2\frac{1}{2}$ inch electric box. The receptacle can be side or rear wired. The face of the unit carries two buttons: a red reset button and a blue test button.

PLUGS[15]

The purpose of a plug (sometimes called a connector), regardless of its shape or the number of its blades or prongs, is to make an electrical connection by physical means to a branch power line. As in the case of other electrical components, there are a large number of styles and types, including older units and a variety of more modern types. Plugs that have had many years of use or that have been mishandled can present an electrical hazard and should be replaced, not with an identical unit, but upgraded with a modern plug. In some instances this may mean replacement of its corresponding receptacle.

Although a plug is a simple device it can lead to trouble in some electrically operated components. If the prongs are too close or too widely separated, the plug may not fit into the openings of the receptacle. If the prongs are loose, there may be no contact or intermittent contact with the receptacle, and so the connected component may not work or may do so sporadically.

The connecting wires to the plug may get warm or may remain cool depending on the current demand of the load. If they get hot, the load may be demanding excessive current, the operating device may be shorted, the connecting wires may be the wrong gauge (a thicker wire may be needed), or there may be a poor connection to the plug. Hot wires or a hot plug are an indication of electrical trouble.

Plug Body

The plug body has several functions. It houses the cord connections and it works to protect the cable connections to the prongs. It is designed (or should be) for easy removal of the plug from a receptacle. The body of the plug should be large enough to permit it to be grasped readily and firmly. Better plugs are serrated or ridged. The worst plug body is made of smooth plastic and is small. A plug should never be removed by pulling on its wires, even though some plugs make it tempting to do so.

Types of Plugs

Sometimes several names are used for the same plug. They can be listed as 120-volt or 240-volt plugs, two-prong types, polarized plugs, round and flat-wire types, replaceable and nonreplaceable, molded and nonmolded, quick connect, right angle, low current and high current, high voltage high current, and adapter plugs. There are many others, as well.

[15](*NEC: Article 410–56 (f); page 370*)

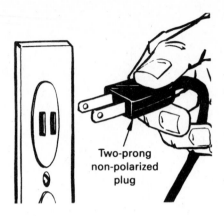

Figure 4–14. Nonpolarized two-prong plug and its accompanying receptacle. The plug is a molded type. Both blades have identical dimensions. Do not use this plug for new work. It is desirable to replace this plug and its receptacle with a three-prong grounded type.

Two-prong non-polarized plug

Basic two-prong plug.   Once one of the most popular plugs, but also one of the least desirable is the basic two-prong, nonpolarized type (Figure 4–14). Commonly, it is molded directly onto a two-wire line or can be supplied as a separate unit.

Electric iron plugs.   One problem often appearing in connection with electrical appliances is that the unit works well for many years and the part that becomes defective is not the appliance but rather its plug. An example is the electric iron plug and quite possibly its associated cord. The cord is a round type packed with asbestos and then covered with a black fabric. The plug, as shown in Figure 4–15, is the type that slides onto a pair of circular prongs extending from the electric iron. The plug can be opened by turning the center screw with a flat-blade screwdriver. The plug then separates into two parts, one of which contains screws for the connecting wire. A spring guard is used to protect the wire covering for a few inches above the plug.

Modern electric irons use molded plugs and have a direct, unreachable wire connection inside the iron.

Flat- and round-wire plugs.   One plug identification method refers to the plug by the shape of the associated wire. These are known as round-wire and flat-wire types. Round-wire plugs are often used in connection with relatively

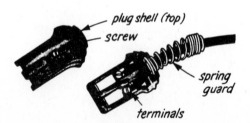

plug shell (top)
screw
spring guard
terminals

Figure 4–15.   Plug used on old electric irons.

high current appliances such as electric irons and some vacuum cleaners, while flat-wire plugs are used with lower-wattage components, including electric shavers, fans, and clocks.

Round-cord plugs. The body of a nonpolarized plug can be either flat or round. The flat plug, as its name implies, is used with wire that has a flat structure, such as a lamp cord. The round plug is intended for use with round cords (Figure 4–16).

Round plugs not only conform to the shape of the connecting wire but are often (but not always) equipped with an extension that consists of a pair of brackets held together with machine screws. When these are tightened, they make a firm grip on the wire and act as a strain relief. The disadvantage of the round plug is that its body uses extra room when plugged into a receptacle, interfering with the insertion of another plug.

Nonpolarized plugs. A nonpolarized plug can be easily identified. It has a pair of equally shaped prongs and can fit into a receptacle in either of two ways. It is called nonpolarized since it does not use a grounding terminal. Nonpolarized plugs can be nonreplaceable units molded directly on a connecting wire or separate units not on wire. The molded plug may be rubber or some synthetic rubber material or plastic. The plastic units are superior since they do not deform and keep the prongs firmly in place, which is sometimes not true of rubber types. Plugs having a flat body are preferable to round types since they permit a better finger grip for plug insertion and removal. However, some rubber plugs have an extension from the plug body and so have the same advantage. Always insert or remove a plug from its receptacle by the plug body or its extension.

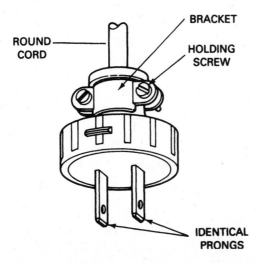

Figure 4–16. Nonpolarized round-cord plug. Plug is held tightly on round cord by bracket.

There are two methods for updating a nonpolarized plug and receptacle wiring. One is to replace the receptacle and plug with polarized units. The other is to identify the ground connection of the receptacle and to mark the plug prong in the same way, possibly with a white paint mark. The hot and neutral wires of the receptacle can be selected by the use of a neon test lamp.

Polarized plugs.

A polarized plug is easily recognizable since one of its blades is wider than the other. It will not fit into a nonpolarized receptacle, and no attempt to force it to do so should be tried. Some polarized plugs have spring-action prongs, such as the one shown in Figure 4–17, ensuring better electrical contact with the receptacle.

The wider of the two prongs is the neutral and is the prong that makes contact with the neutral white wire of the branch wiring. The narrower prong is the hot contact, which makes contact with the black-coded wire of the branch circuit. By applying a dab of paint, you can designate the neutral prong; also apply a similar paint spot on the neutral entrance groove on the receptacle. It is important to remember that the neutral must never be interrupted. Current breaks are always with the black or hot lead.

Right-angle plugs.

Two angles are commonly used by wire coming into a plug. The most common is the on-axis connection, in which the wire makes a 180° angle with the plug; that is, the wire comes straight into the plug. The other is the right-angle plug in which the connecting wire makes a 90° turn, as indicated in Figure 4–18. The advantage of the right-angle plug is that is avoids the need for bending the connecting wire and for making a neater installation should a wire bend be necessary.

Quick-connect plug.

The plugs in Figure 4–18 are quick-connect types that use a pair of sharp metal points to make the electrical contact between the wires of the cord and the plug. The plug is equipped with connectors that pierce the insulation of the wires. This is the quickest method of connecting a plug; there is no need for stripping wires and no chance of a wire strand becoming loose and

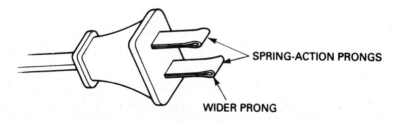

SPRING-ACTION PRONGS

WIDER PRONG

Figure 4–17.   Polarized plug.

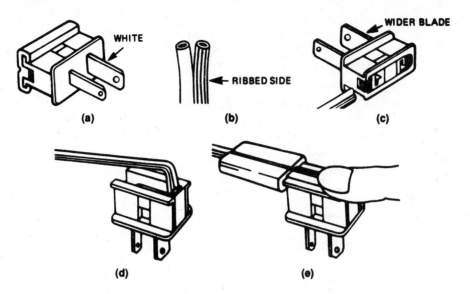

Figure 4–18. Right-angle plugs with automatic cable attachment feature. (*Courtesy Gilbert Manufacturing Co., Inc.*)

presenting the possibility of a short. However, it does make a higher-resistance connection than wrapping a wire around a screw.

Quick-connect plugs can be polarized or nonpolarized, but are only two-prong types. They can be on axis or form a 90° angle as shown in Figure 4–18. Drawing a shows a quick-connect on-axis plug and a section of lamp cord is shown in drawing b. The insulation of the wire is ribbed for easy identification of the hot and cold leads. The wire that connects to the wider blade is the neutral lead. Either of the wires of the lamp cord can be the neutral, but once it is selected it should be used consistently for all plug installations.

The lamp cord is inserted as in drawing c. The neutral wire is on top; the hot lead is below it, and so the neutral will automatically be connected to the wider prong. A sliding tab on the plug is used to force the metal points into the wire.

Drawings d and e show how to connect the wire to a right-angle type of quick-connect plug.

Grounded plug. Sometimes called a three-prong plug, the grounded plug is more desirable than the two-prong types, whether polarized or nonpolarized, for it is used with three-wire cables consisting of a hot lead, a neutral, and a ground wire.

The construction of this plug (Figure 4–19) is such that it can fit into its corresponding receptacle in just one way. Two of the prongs are flat-blade types, while the third is tubular. If the plug is held as indicated in the drawing the right

THREE-PRONG PLUG

Figure 4–19. Three-prong plug and associated receptacle.

prong is the neutral; the left is for the hot lead. As shown, the tubular prong is at the top.

Plug Voltage and Current Ratings

All plugs, whether two prong, polarized or nonpolarized, or three prong, have voltage and current ratings. No appliance may have a higher voltage or current designation than the plug to which it is connected.

As an example a two-prong, two-wire type could be rated at 15 amperes, 120 volts, and could also be polarized or nonpolarized. It might have screw terminals and is intended for use with AWG No. 20, 18, or 16 flat or round cord.

Three-prong plugs are available as 15- and 20-ampere types, 120 volts. It is easy to distinguish between the two. The 15-ampere plug has its two flat blades parallel to each other. The 20-amperes plug has its blades at right angles. These three-blade plugs are illustrated in Figure 4–20.

Special-purpose Plugs

High-current, double-voltage plug. The plug in Figure 4–21 is current rated at 30 amperes and is intended for either 120 or 240 volts. A pair of 120-volt lines is used to obtain the 240 volts. The cable supplying the electrical power is a three-wire type with a black lead supplying 120 volts, a red lead for 240 volts, and a neutral wire. There are two hot-lead blades mounted at angles to each other. The neutral blade is made of two lengths of metal joined to form a right angle. The two hot-lead blades have the same shape and thickness.

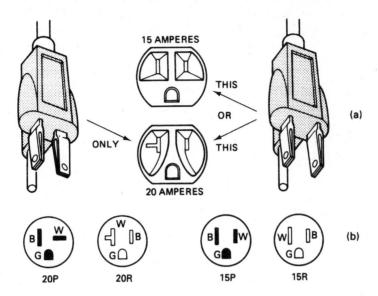

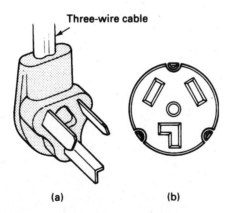

Figure 4–20. (a) Fifteen- and 20-ampere plugs and associated receptacles. (b) Receptacle and plug symbols.

High-current, high-voltage plug. Unlike the preceding plug, this plug is intended for use in connection with a 240-volt power line. It is not meant for use with 120 volts. As shown in Figure 4–22, the two similarly shaped blades have the same length and thickness, with one of these for use on 240 volts, while the other is the neutral. The semicircular blade is the ground connection.

Fifty-ampere, double-voltage plug. The drawing in Figure 4–23 is that of a 50-ampere plug. The centrally located blade is the neutral connection, while the other two prongs are for 120 and 240 volts. All the prongs have the same construction, with the two voltage blades tilted away from each other.

Figure 4–21. (a) Double voltage plug and (b) its receptacle.

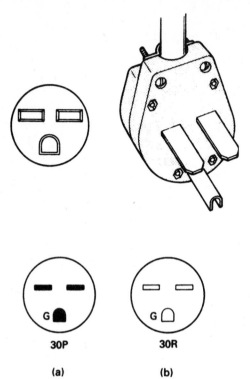

Figure 4–22. (a) 240-volt receptacle and (b) plug. Both have a 30-ampere rating. Symbols are shown below. R, receptacle; P plug.

Adapter plug. It is possible to use a three-blade plug in a receptacle intended for double-blade types with the adapter illustrated in Figure 4–24. This assumes that the electric box holding the receptacle is a metal, not a plastic, type and that the center screw holding the face plate is grounded. The adapter plug is equipped with a pair of equal-sized blades for insertion into the receptacle. The plug has a flexible wire terminating in a spade lug for connection to the center ground screw of the cover plate.

Consider the adapter as a temporary expedient. A better arrangement is to replace the receptacle with one that accepts three-blade plugs, (*NEC: Article 410–58e; page 372*).

Cube-tap plugs. A cube tap is a combination plug and multiple receptacle. The advantage of the cube tap is that it immediately supplies two or three receptacles. The plug, made of a rubberlike material or plastic, is inserted into an active receptacle, and when this is done, its receptacle portions are available as supplementary receptacles.

The cube tap, however, has a number of disadvantages. It looks unattractive since it will have a number of electric cords connected to it. It makes it easy to

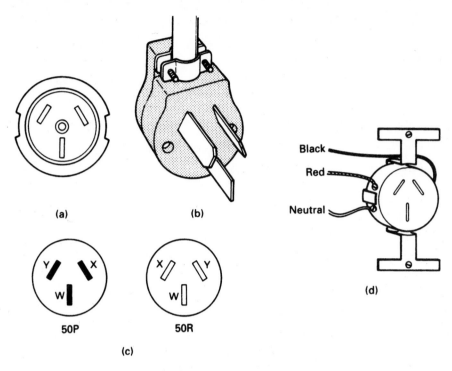

(a) (b)

50P 50R

(c)

Black
Red
Neutral

(d)

Figure 4–23. (a) 50-ampere receptacle and its plug (b). Symbols for the receptacle and its plug. (c) W is the neutral; X and Y are the hot leads. R = receptacle; P = plug. Connections (d).

overload a branch line, depending on the number and wattage requirements of the electrical components plugged into it, and if the cube tap makes poor contact with its receptacle, it may make the plugged in appliances either inoperative or operate poorly. The cube tap may not be used with a dedicated branch line, a line that is intended for use by a single appliance. The fact that a cube tap is used is an indi-

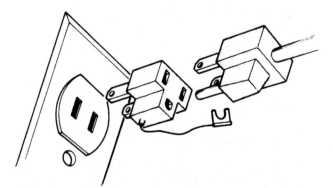

Figure 4–24. Adapter plug to permit three prong polarized plug to use nonpolarized receptacle. Not recommended for permanent use.

cation that there are an inadequate number of receptacles or a possible need for another branch line.

Extension cord cube tap. An extension cord cube tap consists of a length of wire, 3, 6, 9, or 25 feet, and so on, equipped with a single plug at one end and a cube tap at the other. The plug is inserted into a receptacle, with the cube tap acting as supplementary receptacles.

The extension cord cube tap is useful when an electrically operated tool has a power cord that will not reach an existing receptacle. The best plug on the extension cord cube tap is a three-prong type. The supplementary receptacles at the other end of the extension cord should be for three-prong plugs.

Do not regard the extension cord cube tap or the individual cube tap as permanent devices. They are simply indications of inadequate wiring.

Outdoor Plugs

Plugs are manufactured in accordance with their in-use requirements. A plug used indoors will have different environmental needs than one intended for outdoors; and each will be UL tested accordingly. The outdoor plug must be able to withstand temperature changes, dampness, and humidity and meet more stringent UL tests.

The outdoor plug is a three-terminal type having a removable shell to protect the internal connections of the wires. The procedure for replacing the plug is the same as rewiring an indoor three-terminal plug, (*NEC: Article 200–10b; page 51*).

Replacing the outdoor plug. With the plug disconnected from its receptacle, cut the round cable and discard the old plug. Remove the round fiber washer from the front of the new plug by unfastening the screws holding it. The plug cover is held in place by a pair of clamps. Unfasten the screws on either side of the plug clamp. Push the round cable through the center entry of the shell. The round cable will have three wires: white, black, and green. Strip each of these wires back about $1\frac{1}{2}$ inches. Strip about $\frac{3}{4}$ inch of insulation from each wire end and then fasten them to the three screws on the inside of the plug. One screw will be colored silver, one will have a brass finish, and the last will be green.

Wrap one wire under each screw head in a clockwise direction. Connect the white wire to the silver screw, the black wire to the brass screw, and the green wire to the green screw. The wires will probably be stranded types, so twirl each in a clockwise manner, making them as close to a solid conductor as possible.

Once the three wires have been connected, move the plug back into its shell. Tighten both clamp screws. Replace the fiber washer on the front of the plug, holding it in place with screws.

Frequent plug removal.   Removable plugs are available in different forms. If the plug is to be inserted and removed from its receptacle fairly often, the best type for this purpose is the one having a finger extension. This plug is available with two prongs plus a ground connection. The active prongs are flat; the ground connection may be flat or cylindrical. Another type of plug is the twist lock. These have prongs that catch and hold when the plug is inserted and twisted. Unlike other plugs, the twist lock is equipped with an exit clamp equipped with screws, a device that grips and holds the connecting wires.

RECEPTACLES[16]

A receptacle (sometimes called an outlet), is a device for permitting the easy connection of a load to a branch power line. Receptacles have no moving parts and so rarely cause electrical problems, but these can exist. Like any other electrical component, they sometimes need updating.

The voltage and current supplied to a load must pass through a receptacle, and so the receptacle must be able to meet the power demands of the load. All receptacles are wired in parallel with other receptacles connected to a specific power branch. The voltage supplied by that branch is always present at the receptacle, and so all receptacles are live except for those that are switch controlled. (*NEC: Article 210–52a; page 63*).

The receptacle is the controlling factor in the design of a plug. The prongs or the blades of the plug must fit snugly into the slots of its associated receptacle.

Receptacles can be categorized in a number of ways. A basic listing would be indoor receptacles, outdoor receptacles, and switched receptacles. Another classification would be to designate them as in wall or on wall. The most common receptacle is one that is gradually being replaced (or should be) and consists of two slots having identical dimensions permitting the insertion of a plug whose blades are identical, (*NEC: Article 200–10b; page 51*).

Identifying Receptacles

The function of a receptacle is to accept a plug and act as an intermediary for the delivery of electrical power to a component. The receptacle can be mounted in an in-wall electrical box or be a wall-surface mount type, (*NEC: Article 100; page 24*).

The purpose of a receptacle is to act as a terminal for the two or three wires of a branch power line. These wires form a link with a fuse or circuit breaker box,

[16](*NEC: Article 410–57; page 371*)

and the receptacle can be inserted at any convenient and accessible point anywhere along the branch line (*NEC: Article 210–21; page 60*).

It is necessary to identify receptacles since there are so many different types. First, they are known as single, duplex or double, and triplex or triple. A single receptacle can accept a single plug only, a duplex two plugs, and a triplex three. Receptacles can also be identified by their voltage and current ratings, by the type of plugs they use, or by the way in which they can be wired. Some receptacles can only be wired by side-positioned screws and so are known as side-wiring types; some are wired at the back and are called back wired. Some receptacles are combination units consisting of a switch and a single receptacle.

Switched receptacle.

Most receptacles are permanently active types with branch line voltages always available. However, a switch can be substituted for one of the receptacles in a duplex-receptacle arrangement. The disadvantage is the loss of one receptacle, but in exchange the remaining receptacle is controlled and can be turned on or off. This is helpful when using electrical components that do not have a built-in switch. Another advantage is that the switch can exercise control from some distance.

The switch is ordinarily a single-pole, single-throw type rated at 10 amperes, 120 volts or 5 amperes, 240 volts. The single receptacle is specified at 15 amperes, 120 volts. The load must not exceed the switch rating.

Dedicated receptacle.

A dedicated receptacle is one that always has a single, specific application. For example, it may be used only for operating a photocopy machine or for supplying power to a particular motor.

Two-prong nonpolarized receptacle.

The two-prong type was once the most common receptacle, but it is being replaced by the three-blade type. The two-prong receptacle, shown in Figure 4–25, is intended for the insertion of a nonpolarized two-prong plug.

Figure 4–25 is a front view of this receptacle. The plug's prongs are inserted in the entry slots. These are of equal dimensions and so the plug can be put into the receptacle in either of two ways. This receptacle can accommodate two plugs.

Figure 4–25.   Nonpolarized receptacle.

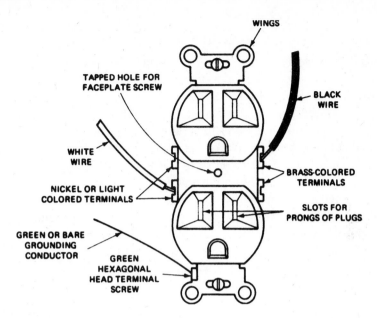

Figure 4–26. Front view of a three-prong polarized duplex receptacle.

Polarized plugs have a pair of prongs. The width of one of the blades is such that it prevents pushing the plug into this receptacle. It is inadvisable to alter the blade to permit its use.

The wires of the branch line can be connected to a pair of screws, with the black lead wrapped around one screw; the white lead is attached to another screw on the opposite side of the receptacle. Some of these receptacles may have a screw for the connection of a ground wire. As an alternative, the receptacle may have holes in the rear for a push-in connection.

Duplex, polarized, three-prong receptacle. Figure 4–26 shows a front view of a duplex polarized receptacle. The unit has a pair of metal extensions, known as wings or ears, that are used to fit the unit against plasterboard or dry wall (Figure 4–27). Their purpose is to help hold the receptacle in place. No

Figure 4–27. Ears are used to help secure the receptacle in position.

screws are used or are necessary. There is also a tapped hole in the center of the receptacle to accommodate a $\frac{1}{60} \times \frac{1}{4}$ inch machine screw. This screw is used to hold a cover plate in position. If the receptacle is to be mounted in a metal electric box, the screw can be used as a ground connection for external equipment, (*NEC: Article 410–56e; page 370*).

The receptacle in Figure 4–26 is designed to accommodate three connecting wires: a black or hot lead shown at the right, a neutral white wire, and a ground lead at the bottom, (*NEC: Article 410–58b; page 372*). The ground lead can be a bare wire, but if it is insulated, it will be coded green. With some receptacles a grounding clip is used instead, and the ground wire is slipped under it and held in place against the metal by spring action. Plastic electric boxes do not have this feature. The holding screw for the white wire will have either a nickel finish or a light color such as silver, while a similar screw for the black lead will be brass colored. For screw connections, wrap the wire in a clockwise direction. There is a pair of screws on both sides. These screws are captive types (Figure 4–28), which limits the amount by which they can be unscrewed, making it impossible to drop them accidentally.

You will find a pair of push-in holes on the rear of the receptacle. All that is necessary is to strip the black and white leads and insert them. But whether using the insert holes or the connecting screws, be sure not to transpose the wires. Doing so will result in a short circuit. Pull on the wires to make certain they are firmly connected.

When using the push-in holes, it is necessary to strip the connecting wires by the correct amount. On some receptacles you will find a strip gauge indicating the amount of insulation to be removed from the connecting wires. The wires must be stripped enough to make good electrical contact. If stripped too much, some bare portion of the wire will be exposed. Receptacles with push-in holes are also equipped with a release mechanism. This consists of a pair of slots. Insert a bare solid wire or the blade of a small screwdriver while simultaneously pulling on the connecting wire to be removed.

There are a total of six entry slots on the front of this receptacle, as indicated in Figure 4–26. The semicircular slots are for the insertion of the ground prong, and the two vertical, parallel slots are for the neutral and hot prongs of the plugs. The slots at the right are smaller than those at the left. These slots are connected

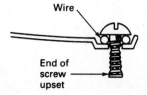

Figure 4–28. The upset prevents the screw from being completely removed. Such screws are sometimes called captive.

internally to the hot, brass-colored terminals of the receptacle. The other two, larger slots are the neutral connections.

Useful data supplied with this component are its voltage and current ratings. It should also carry the Underwriters' Laboratories® logo.

In-wall, recessed, and on-wall receptacle.

Most receptacles are in-wall types; that is, they fit into an electric box that is in-wall mounted. The front of the receptacle is then flush with the outside surface of the wall. For some applications, such as a kitchen clock, the receptacle is mounted in a recessed electric box. This permits positioning the clock directly against the wall surface. There is also room to accommodate any extra wire leading to the clock.

Another type is the surface mount. The electric box and its receptacle are positioned directly on the wall. Since the connecting wires are above the wall's surface, they are covered by a metal channel that is fastened to the wall.

Isolated ground receptacle.

When mounting a conventional receptacle in a steel box, the ground is commonly established through the existing electrical system. This is done either by using the grounding clip on the receptacle's mounting strap or by running a ground wire, which is part of the existing system, to the grounding screw. Thus, even when a separate wire is brought into the receptacle, it is still tied into the normal ground. This occurs since the mounting strap is in contact with the box grounding system.

The conventional grounding receptacle provides safety for personnel and equipment. However, the ground network also serves as a giant antenna and conductor of electrical noise. This electrical noise is electromagnetic interference and is caused by numerous transient ground currents. This can produce random electrical signals in the grounding system.

As a result, sensitive electronic equipment such as point-of-purchase terminals, accounting machines, computers, and highly sensitive medical and communications equipment can pick up these transient signals, which can interfere with the proper operation of the equipment. Figure 4–29a shows the wiring of a typical conventional receptacle.

The solution to this problem is the isolated ground receptacle. Its arrangement is illustrated in Figure 4–29b. It is similar to a conventional receptacle except for one important change. An insulating barrier construction isolates the ground contacts from the mounting strap. The grounding screw is connected directly to the grounding contacts. In this way, the ground contacts are separated from the mounting strap and also from the conventional grounding system. The isolated ground circuit is completed by running a ground wire to the ground screw. This ground wire comes from a separately derived ground system in accordance with *NEC: Article 250–74, exception #4, page 141.*

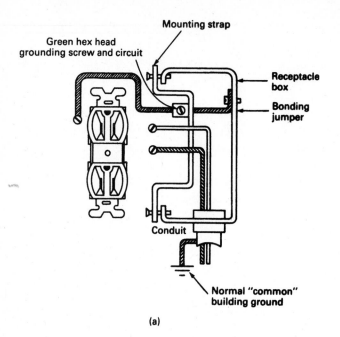

(a)

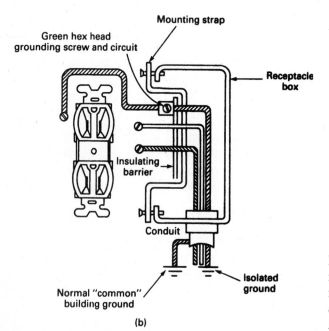

(b)

Figure 4–29. (a) Conventional receptacle and (b) isolated ground receptacle. (*Courtesy Hubbell Incorporated, Wiring Device Division*)

The isolated ground, sometimes called a *pure ground,* can be kept relatively free of electrical noise. This is achieved since the grounding network is shorter, has fewer sources of noise, and is connected to the earth ground at a single point. Using this isolated grounding system, electronic equipment can be kept operational and free from electrical interference.

Isolated ground receptacles can be identified by the use of an orange-colored wall plate carrying the designation "Isolated Ground." These are smooth nonconductive nylon wall plates with the lettering hot stamped in black.

Hot receptacle. If the electric cord of an appliance plugged into a receptacle feels warm to the touch, wrap the portion of the cord near the receptacle with a dry towel, and pull the plug out. If there is evidence of burning or sparking or smoke is coming out of the receptacle, do not touch it. Instead, shut off power to the receptacle by opening its circuit breaker or removing its fuse. The problem may be in the receptacle or in the appliance.

Line cord receptacle. A line cord receptacle consists of a length of extension cord with a receptacle at one end and a plug at the other, with both usually molded types. The cord comes in various lengths of 3, 6, 9, and 12 feet. Older cords used nonpolarized plugs; modern cords are equipped with two-prong polarized plugs. The connecting line cord is a flat type.

Cords are also available for carrying larger amounts of current; these use round wires. In some instances the cord has a test lamp at one end with a wire guard to keep you from accidentally touching the hot electric light. In some cases the base of the light contains a receptacle. The plug supplying input power to the cord is a grounded type with a pair of parallel blades and a semicircular prong for a ground connection.

Some line cord receptacles are supplied in a housing with two or more built into the face of the housing. The cord is a round type equipped with a molded, three-blade, polarized plug. The advantage of this setup is that the arrangement is always regarded as temporary, and the cord is wound back into its container when finished. The container has a handle so it can be easily carried from one job to the next. The use of a case helps minimize the possibility of tripping over the line cord.

Receptacle Adapter

A receptacle adapter is used for increasing the number of receptacles from two to as many as four. The rear of the adapter is equipped with blades for insertion into an existing baseboard-type receptacle pair.

Hospital-grade Receptacle

Hospital-grade receptacles outwardly resemble the type commonly used in homes but are constructed to more rigorous specifications. Underwriters' Laboratories verifies that wiring devices with a hospital-grade listing have passed a series of "torture tests," supplying assurance of continuous operation despite high-abuse usage.

All exposed metal parts are corrosion resistant to assure trouble-free performance even in harsh environments with exposure to corrosive liquids such as cleaning fluids, rug shampoos, rain, chlorinated water sprays, salt air, or chemical fumes. Hospital-grade receptacles (and plugs) withstand impact wear, crushing, and corrosive conditions that would destroy ordinary heavy-duty units.

Two types are available in standard colors and are rated at 15 amperes, 120 volts and 20 amperes, 120 volts. The wall plates for these receptacles are all-nylon composition for maximum resistance to impact, grease, oils, moisture, and cleaning fluids.

Hospital-grade receptacles also offer true isolated grounding. A separate pigtail grounding lead on the receptacle strap provides a pure grounding path that is separate from the normal grounding circuit. Conventional grounding interference, which may cause sensitive electronic equipment such as computers, cash registers, and medical equipment to malfunction, is eliminated. The separate grounding path eliminates electromagnetic interference.

Limitations of the Two-hole Receptacle

Many modern appliances, including shop power tools and electrically operated kitchen devices, are equipped with a three-wire power cord ending in a three-prong plug. The three wires in the power cord consist of the hot or black lead, the white or neutral lead, and a wire that is bare or covered with green insulation. This is the ground wire. Of the three wires, two must be continuous conductors, and these are the ground and neutral wires. The ground and the neutral wires are essentially connected in parallel. The electric current circulates through these three wires. The amount of current flow through the black lead should be equal to the sum of the currents flowing through the ground and neutral wires.

An electrical pressure, known as line voltage, exists between the hot wire and the ground and neutral wires. The ground lead connects the outer metal housing of appliances to the third prong of the plug. The act of connecting a three-prong plug into a properly wired three-hole receptacle automatically grounds the metal frame of an appliance and any metal parts connected to it.

Receptacle Power Ratings[17]

The usual in-home receptacle is rated to handle a maximum of 15 amperes. To calculate its power-handling ability, multiply the existing line voltage by the maximum current. Assuming a line voltage of 120, multiply it by the maximum current rating, in this case 15 amperes. $120 \times 15 = 1800$ watts or 1.8 kilowatts.

This is also the maximum power to be handled by the branch line. If that line has five receptacles, they will distribute the load, and the amount of that load for a particular receptacle will depend on the wattage required by the plugged-in load. Thus, whether a receptacle will be overloaded or not depends on the connected appliance (or appliances in the case of a duplex receptacle).

Disadvantages of Receptacles and Cube Taps

Receptacles are not only useful but are necessary as a quick and easy means for connecting appliances into a power line. However, the prongs of a plug on an appliance make only sliding contact with the metal elements in the receptacle. Yet current must pass from those metal contacts in the receptacle to the prongs of the connected plug and then into the appliance. This is the weak link in the electrical system, for elsewhere in the system the connections are extremely tight, with wires wrapped around each other and the connection made more secure with a wire nut, or else wrapped around a screw and tightened by a nut.

Several things can be done to overcome the inherent weakness of the receptacle-plug arrangement.

1. Make sure that the portion of the receptacle receiving and housing the plug extends beyond its cover plate. In other words, the prongs of the plug must go as far into the receptacle as possible, with the end of the plug firmly up against the receptacle.
2. The plug-receptacle connection must be tight. Tug gently on the wire connected to the plug. If the plug falls out of the receptacle, the fit is poor. Spread the prongs of the plug slightly or twist them using two gas pliers. However, bending and twisting the prongs sometimes loosens the prong supports. In that case it is better to replace the plug with a new one. Don't keep the old plug for possible use at some later date. Throw it away.
3. The prongs of a plug are mounted in a rubber substitute or plastic. After

[17]*(NEC: Article 410–56; page 370)*

some use, one or more of the prongs can work their way loose. If they can be wiggled, replaced the plug with a new one.

4. After a plug has been working, remove it and touch the prongs. They may be mildly warm. If more than that or if hot to the touch, the plug is making poor contact. This is more noticeable with current-hungry appliances than with those that require small currents.

5. Always check the power cord connected to the plug. If it has cracks, has been painted over, or is too short, it is advisable to replace it with a cord that has a nonreplaceable plug. The plug must be a type that fits the associated receptacle. If the receptacle is a two-prong nonpolarized type, it would be well at this time to replace it with a three-slot polarized unit.

Testing Receptacles

If an appliance does not work when plugged into a receptacle, the problem can be in the receptacle, the branch wiring, or the fuse or circuit breaker. To narrow the location of the fault, try the appliance in a different receptacle. If it works, the problem can be in the service box, the branch wiring, or the receptacle. It can also be in some preceding receptacle from which the inoperative receptacle gets its power.

There are various ways of testing receptacles, and the procedure described is just one. One method is to use a neon glow test lamp, an inexpensive device consisting of a small test bulb connected to a pair of insulated wires ending in metal prongs that can be inserted into the openings of a receptacle.

To use the neon tester, insert the prongs into the receptacle slots. If the bulb lights, the receptacle is functioning. If the receptacle is a duplex type, and this is usually the case, test both. It is possible for one to function and the other to be defective.

If the glow lamp does not light, touch the center screw of the cover plate with one test lead, and insert the other test lead into the slots of the receptacle, one after the other. It should light in one position, but not in the other. The center screw must be free of paint. The slot that produces no light is the ground connection.

If the two-prong receptacle is a polarized type, use the same test. Use it also for a three-prong polarized receptacle. The neon lamp should glow when testing from the hot to either the ground or neutral openings.

There are several other alternative test procedures. Use a lamp known to be in good working order to test the receptacle or a lamp-equipped extension cord. Or use a multitester set to read 120 volts ac.

If testing indicates that the receptacle is not active, although all other receptacles and lights are functioning, then open the circuit breaker or remove the fuse for the associated branch line, remove the receptacle cover plate, and then remove the receptacle. The receptacle may be defective or may have a loose connection.

Twist-lock Receptacles

A plug should fit snugly in its receptacle. If it is loose, it can cause operating failure of the connected component or intermittent operation. The problem may be cured by bending one of the blades slightly outward. This is more easily done with receptacles equipped with a plastic body. If the body is rubber, it may be too yielding to permit a permanent prong bend.

Other than bending the plug's prong, there are several alternatives. One is to replace the plug. If it is a removable type, use a replacement. If molded, cut the connecting wires near the plug, strip them, and use a replacement plug. Another technique is to use a new line cord equipped with a molded plug. It will be necessary to use a nonpolarized or polarized plug, depending on the requirements of the receptacle.

Still another solution is to use a twist-lock arrangement (Figure 4–30). This means using both a new receptacle and a new plug, both of which are twist-lock types. The technique involves inserting the plug and twisting it somewhat to lock it into position, an action that results in positive contact between the plug and its receptacle. An installation of this kind may be required where a wall is subject to vibration, possibly from a nearby motor.

Recessed Receptacles

After a plug is put into a receptacle, the connecting wires to the plug extend away from the wall holding the receptacle. For many applications, this is satisfactory and there is no reason why the wire should hug the wall. But there are times when it is desirable to have the plug cord as close against a wall as possible. An electric kitchen clock is one example. In this instance, a recessed type of receptacle is preferable.

Figure 4–30. Types of twist-lock receptacles.

Combined Switch and Receptacle

Instead of a duplex receptacle, it is possible to use a combined switch and receptacle. There are two basic types. In one the receptacle is always active and is independent of the switch. The second type uses the switch to control the single receptacle, and the receptacle is active only when the switch is in its closed position. The receptacle is referred to as a switched type.

Safety Inserts

Receptacles are often mounted near the baseboard of a wall and so are well within the reach of children. While the slots of the receptacle will not permit exploration by fingers, children can insert slim metal objects and so become subject to electrical shock.

To prevent this possibility, close all unused receptacles with plastic safety inserts. These should fit tightly into the receptacle so they are not easy to remove by children. The inserts have the added advantage that they eliminate the possible movement of cold air from behind the walls through the receptacle slots, a wintertime benefit.

Surge-suppression Receptacles[18]

Branch power lines can pick up transient, unwanted voltages in two ways. Acting somewhat as an antenna, they are capable of picking up stray magnetic fields. They can also receive transient voltages from equipment connected to their receptacles. These transients can have voltage levels far greater than normal line voltage and can ride in with electrical currents flowing to equipment connected to the various receptacles. A transient, for example, can be developed when the current flow through a coil is suddenly interrupted. In many instances, these transients do no damage, but they are quite capable of incapacitating computers, word processors, and disk drives.

There are two basic types of surge-suppression receptacles, the plug-in type and the receptacle type. The surge receptacle is more desirable since it is less conspicuous and has a larger margin of safety. Two variations are available: single boss, corresponding to a single receptacle, and double boss, in place of a duplex receptacle. The double suppressor (Figure 4–31) is equipped with a power-on indicator light whose glow is reassurance that the receptacle is live and that its surge-suppression feature is functioning. For this reason it is advisable to have the receptacle wall mounted where the light is visible. Failure of the light is an indi-

[18]*(NEC: Article 280A; page 156)*

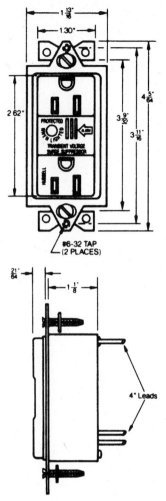

Figure 4–31. Duplex polarized surge-suppression receptacle. (*Courtesy Hubbell Incorporated, Wiring Device Division*)

cation that the surge suppression is not working even though the receptacle itself may be supplying power.

In addition to the warning light, the duplex surge-suppressor receptacle is equipped with a damage alert beeper that continues its warning until the receptacle is replaced. In addition to protecting against line voltage surges, the unit also works to minimize or prevent radio frequency interference (RFI). As indicated earlier, the branch power lines can work as antennas, picking up unwanted transmitted signals and routing them via the power line into connected components.

Surge-suppression receptacles can be installed to retrofit ordinary receptacles and have similar wiring. To emphasize that a surge-suppression receptacle is being used, it can be covered with a distinctively colored wall plate, (*NEC: Article 280–11; page 157*).

There are differences between a voltage surge and a voltage spike; these consist of time duration and amplitude. The voltage spike can exist for a part of a second, but its strength is measured in kilovolts, possibly reaching as high as 6 kV. The voltage surge lasts longer, but its amplitude is less, somewhat on the order of 2 kV (Figure 4–32). Both are capable of damaging equipment via the power line.

Electrical noise via the power line can also cripple sensitive electronic equipment. Known as electromagnetic frequency interference (EMI), it is caused by feedback into power lines by motors and fluorescent fixtures.

As a summary, then, while it is desirable to have only pure sine-wave ac, that sine wave can be accompanied by RFI (radio frequency interference), EMI (electromagnetic frequency interference), and voltage surges.

Some receptacles have built-in devices to supply protection against all three, which are referred to as transients. Transients can have very high voltage peaks and so can damage components plugged into the receptacles of a branch line. The components need not necessarily be turned on since the transient peak voltage may be much higher than the operating voltage of a switch. Transients can cause substantial damage to connected equipment. In the case of computers, transients can cause permanent damage to the memory, resulting in loss of data. They can also harm disk drives.

Multiple Receptacles

The multiple-receptacle unit (Figure 4–33) consists of a rectangular metal box containing from four to eight receptacles mounted in line adjacent to each other. It usually contains an on-off switch simultaneously controlling the on-off condi-

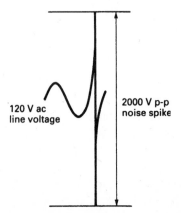

120 V ac
line voltage

2000 V p-p
noise spike

Figure 4–32. High-voltage electrical noise spike superimposed on line voltage (not drawn to scale).

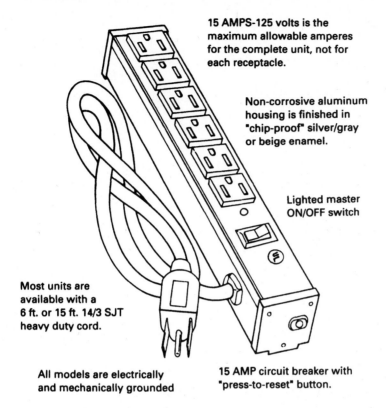

15 AMPS-125 volts is the maximum allowable amperes for the complete unit, not for each receptacle.

Non-corrosive aluminum housing is finished in "chip-proof" silver/gray or beige enamel.

Lighted master ON/OFF switch

Most units are available with a 6 ft. or 15 ft. 14/3 SJT heavy duty cord.

All models are electrically and mechanically grounded

15 AMP circuit breaker with "press-to-reset" button.

Figure 4–33. Multiple receptacle strip. (*Courtesy Brooks Electronics Manufacturing Corporation, Inc.*)

tion of all units that are plugged in. It may be equipped with an indicator light and can also have a circuit breaker.

The multiple receptacle is equipped with a line cord terminating in a three-prong plug that is plugged into a branch receptacle. The advantage is that it supplies the convenience of a number of receptacles, as, for example, those needed at a workbench. However, it becomes very easy to overload the branch line to which it is connected.

Multiple tap. This is a simplified form of the multiple receptacle. It consists of two or three receptacles molded together and equipped with a two- or three-prong plug for insertion into a receptacle. It supplies additional outlets when these are needed, but, as in the case of all multiple receptacles, it can lead to line overloading.

A variation of the multiple tap is one that is equipped with an extension cord sometimes terminating in a two-prong plug. Since the cord must carry the total current of all plugged-in and operating appliances, it should be current rated accordingly.

Multiple-receptacle strips should carry a label indicating the operating voltage and the maximum load current, for example, 15 amperes, 120 volts ac. All the receptacles in the strip are wired in parallel, so all supply the same voltage. The current out of any receptacle depends on the load and can vary from one individual receptacle to the next. The total load is the sum of all the currents from each receptacle.

While the strip is protected at the main box by a fuse or a circuit breaker, it is desirable for the multiple-receptacle strip to carry its own breaker. This helps limit the search for a fault in the event the breaker opens. Another feature is an on-off switch and a red glow lamp. All the receptacles are designed for three-prong plugs, but can accommodate nonpolarized two-prong types. However, this does not make these plugs grounding types. The multiple-receptacle strip is connected to a three-prong type of receptacle with a three-prong plug via a round wire.

Features of receptacle strips.   Not all receptacle strips have the same features; consequently, some may be more desirable than others. When low cost is important, select commercial-grade units. This type may be equipped with a 15-ampere press-to-reset circuit breaker, a connecting line cord having a length of 6 to 15 feet, and a lighted on-off master switch.

Some strips are equipped with transient surge and noise suppressors. These strips can protect equipment such as electronic cash registers, minicomputers, microprocessors, hospital equipment, and other electronic components sensitive to power-line surges. These electronic devices minimize the possibility of lost data in computer memory from transient spikes and reduce errors from small voltage spikes seen as input signals by solid-state devices.

Surge suppressors offer surge current protection of 6500 amperes, a clamping surge voltage of 325 volts, and a clamping response time of 15 nanoseconds (a nanosecond is one-billionth of a second). While strips are designed for permanent plug in to a constantly active receptacle, some are equipped with a remote operating switch.

Strips are supplied with as few as four receptacles to as many as 20. The strip may be equipped with a combination switch-pilot light to indicate its on or off condition. Some are also designed for heavy industrial equipment use.

It is also possible to obtain a strip with each receptacle controlled by its own on-off switch plus a master switch that permits turning all the receptacles on or off simultaneously.

There are medical receptacle strips listed as hospital grade, designed for hospitals, laboratories, or any environment that requires hospital-grade components. However, when used in a hospital environment, grounding reliability can only be achieved when the hospital strip is plugged into a receptacle marked "hospital grade" or "hospital only." This strip can also be used on instrument carts and in scientific work areas.

Workbench receptacles.   Electricians may find it helpful to have a caster-mounted workbench permitting instant power distribution availability. The bench is equipped with six receptacles, can be supplied with a circuit breaker with a typical rating of 15 amperes, 120 volts, and uses a 14/3 (14-gauge wire, three conductors) power line connector having a length of 6 to 15 feet. The unit can be designed for both surge and noise suppression. The frequency range of noise suppression is typically 100 kilohertz to 300 megahertz, with an attenuation voltage ratio of 10 to 40 decibels or more. The unit may be supplied with an on-off switch and light and may have a 15-ampere maximum push-to-reset circuit breaker. The specifications for the unit should indicate the maximum surge voltage, often 6 kilovolts.

Commercial-type multiple receptacles.   For commercial use, it may be necessary to select an industrial-grade surge and noise suppressor for large office and industrial applications where factory machinery and heavy usage of office, medical, and test equipment and laser scanning systems may cause surge and noise problems in the power line. Sensitive computer systems can experience loss of stored data, false input-output or system failure because of power-line problems. Some units are equipped with an audible alarm system if the surge protection feature is not functioning.

While strips are commonly equipped with six receptacles, those having just two and as many as twelve are also available.

Voltage and current ratings.   The voltage and current ratings of the individual receptacles in a strip are typically 120 volts and 15 amperes. When plugged into a convenient receptacle, they form part of a branch line. Any component plugged into any of the outlets is a load on that branch line, and although its appearance is different from that of a cube tap, the strip presents the same hazard, that of overloading the branch. Unlike the cube tape, however, the strip is often equipped with a push-to-reset circuit breaker. The circuit breaker is at the point of entrance to the multiple outlet and so acts as a breaker for each individual receptacle. The line cord leading into the multiple outlet is typically 4 feet of No. 14/3 SJT using a grounded plug.

The unit is designed for indoor use only. Do not install it near water or where it will be exposed to moisture.

Foreign Receptacles

In some instances it may be desirable to travel in foreign countries with small appliances such as hair setters, hair curlers, a heating pad, or an electric razor. However, these cannot be used abroad for two reasons: the line voltage is 220/240 volts and receptacles will not accept plugs intended for U.S. use.

Figure 4–34 shows the receptacle entry ports in (a) England, Africa, and Hong Kong; (b) Germany and England; (c) Germany, Austria, and Switzerland (recessed receptacles); and (d) South American and Caribbean nations. Insert your appliance into a converter and then plug the unit into the foreign receptacle.

In some instances the wall receptacle may not accept the converter. In that case, bring your problem to the hotel desk. They should be able to supply an adapter.

There are many precautions to follow. Converters have a listed power capacity, some value such as 1600 watts. The appliance you intend using should not exceed this rating. Check its wattage rating before using it. Do not leave the appliance connected indefinitely but remove it after usage is completed. Some countries have a line frequency of 50 hertz and so the appliance may run a bit slower than normal. Sometime you may hear a low humming sound from the converter when it is being used. This is normal. The converter should have a built-in circuit breaker for protecting the appliance. The converter also has a built-in voltage step-down transformer and may become slightly warm.

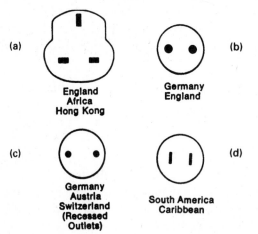

Figure 4–34. Slot positioning for foreign receptacles.

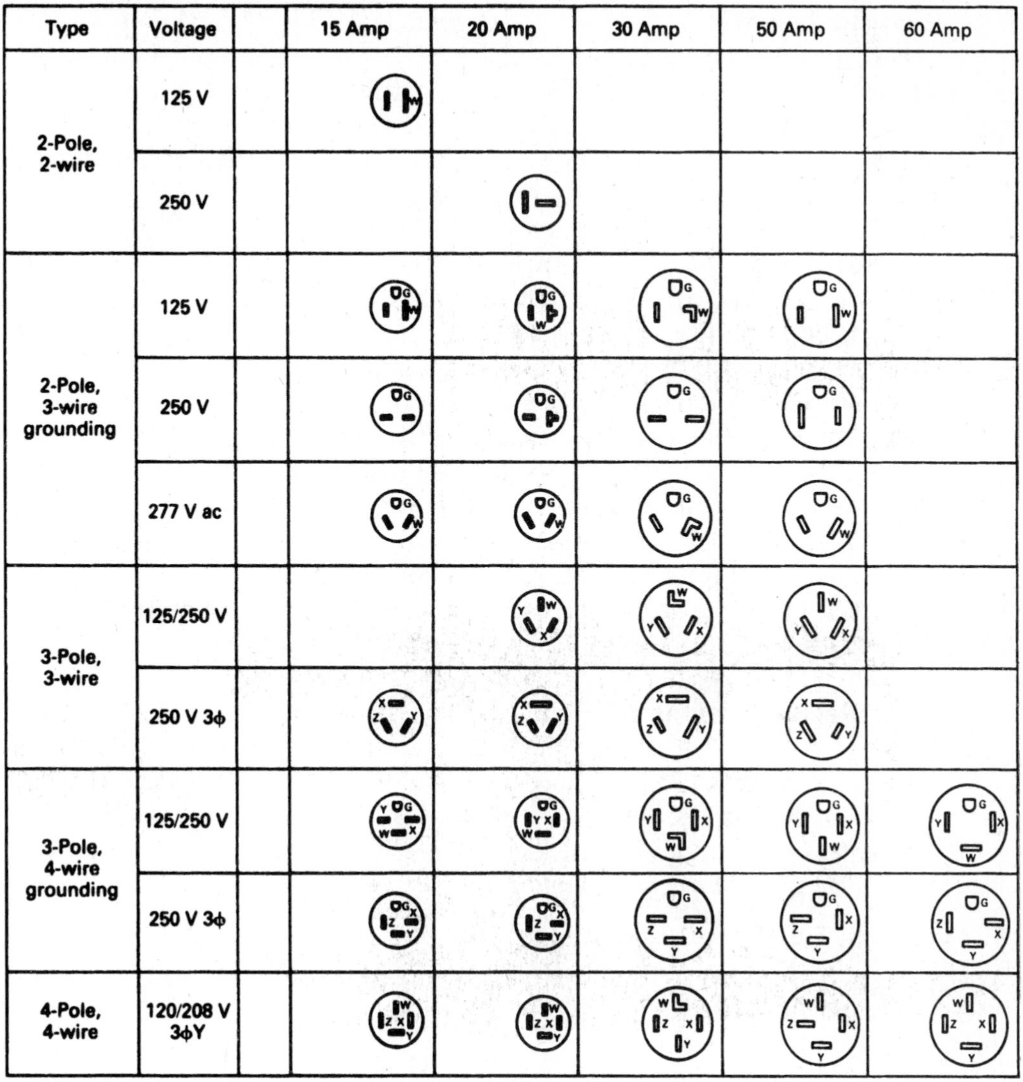

Type	Voltage		15 Amp	20 Amp	30 Amp	50 Amp	60 Amp
2-Pole, 2-wire	125 V		⊘				
	250 V			⊘			
2-Pole, 3-wire grounding	125 V		⊘	⊘	⊘	⊘	
	250 V		⊘	⊘	⊘	⊘	
	277 V ac		⊘	⊘	⊘	⊘	
3-Pole, 3-wire	125/250 V			⊘	⊘	⊘	
	250 V 3φ		⊘	⊘	⊘	⊘	
3-Pole, 4-wire grounding	125/250 V		⊘	⊘	⊘	⊘	⊘
	250 V 3φ		⊘	⊘	⊘	⊘	⊘
4-Pole, 4-wire	120/208 V 3φY		⊘	⊘	⊘	⊘	⊘

Figure 4–35. Receptacle diagrams. (*Courtesy Leviton Manufacturing Co., Inc.*)

Special Receptacles[19]

Ordinarily, you would expect to find a receptacle in the wall, either near a base-board or somewhat higher up. However, it is possible to find them in unexpected places. They are sometimes supplied with furniture, such as a wall unit. Often they are part of the rear apron in high-fidelity sound units, especially in a dedicated system. In such a system, instead of being integrated in a single enclosure, the high-fidelity units, such as preamplifiers, power amplifiers, tuners, cassette decks, and compact disc players, are individual components, with each requiring a connection to a line voltage source. Often, there are not enough receptacles, leading to the use of cube taps, and cube taps inserted into cube taps.

High-fidelity components can be equipped with two types of receptacles, active and switched. The active types are always live and behave in the same manner as a wall-installed receptacle. The switched-type component must be turned on to activate the receptacle.

The voltage rating of these receptacles is always 120 volts. They are never supplied with 240 volts. They may be polarized or nonpolarized. It is important to learn their wattage rating or the amount of current they can supply. This is generally supplied in a separate specification sheet.

Receptacle Diagrams

Receptacles are made to accommodate the various plugs described in the preceding part of this chapter. The chart in Figure 4–35 (see page 135) indicates the various types of receptacles and the voltages and currents with which they are associated. These are all polarized units.

[19](*NEC: Article 210–52; page 63*)

Chapter 5

Electrical Components

PART II

*Switches, Electric Boxes, Cover Plates, Sockets,
and Electrical Hardware*

SWITCHES[1]

Primarily, the function of a switch is to connect conductors, components, or an arrangement of these so that an electric current can flow from one to the other or to discontinue the flow of such a current. Not only one but a number of currents can be so controlled. Switches can also be used as routing devices permitting a current, or currents, to move along previously designated paths.

A switch is not a load. A loss of voltage, a voltage drop across a switch, should be as close to zero as possible. Switches are always wired in series with the hot lead of a wiring system, but are never connected so as to interrupt the ground or neutral leads. Switches are never connected directly in shunt with any power line, that is, from a white lead to a black lead. However, a switch can be

[1]*(NEC®: Article 380; page 320)*

137

wired in series with a load and then the combination can be shunted across the power line, as indicated in Figure 5–1.

If the flow of current is to do useful work, it must be controlled. There are a number of different kinds of current control, but a switch is the most widely used. Basically, a switch is a current-on, current-off device, a go, no-go mechanism. For the most part, switches are manually operated, but some can be light or sound activated.

Commonly, switches are toggle operated; a small handle must be moved up or down or less often, side to side. The switch is usually turned on when the toggle is up or left off when the toggle is down or right. Some are lighted when the switch is in its off position, others when the switch is on. The type of switch to use is determined by its operating conditions, that is, to serve specific functions and meet various convenience factors, environmental requirements, and safety needs, and the number of circuits to be controlled. The choice of a switch can also be determined by whether it is to be flush with a wall, surface mounted, or attached to flexible wires, referred to as cords.

The arrangement shown in Figure 5–1 can be used to operate each lamp independently. Each lamp and its switch are wired in series with each other and the combination is connected in shunt across the power line.

Switch Types[2]

Switches are available in a large variety of types, but all can be arranged under three types of headings: dc, ac, and ac/dc. AC switches are intended for use with alternating current only, dc for direct current only, and ac/dc for either kind of current. Most switches are ac types since ac is so much more prevalent. These switches have certain advantages: they are smaller than dc switches but have equal power capabilities.

DC is commonly used for portable devices that are battery operated. AC switches find their greatest application with components that receive their power from an electric utility.

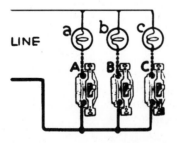

LINE

Figure 5–1. Individual lamp control by using separate switches.

[2](*NEC: Article 380–2; page 320*)

Under the main heading of ac or ac/dc switches, these components can be designated by their pole quantity and the way in which these poles are moved.

Single pole, single throw (SPST)
Single pole, double throw (SPDT)
Double pole, single throw (DPST)
Double pole, double throw (DPDT)

Switches can also be described by physical characteristic or by use. Some types are:

Centrifugal	Heavy duty
Push button	One way
Line cord	Three way
Toggle	Press action
Illuminated	Rocker
Locking	Snap
Dimmer	Key operated
Momentary contact	Open
Motor	Enclosed

Often a switch may be described by using several terms, for example, illuminated toggle SPDT switch.

Single-pole, single-throw switch.[3]   Switches may be open so that their working parts are exposed as indicated in Figure 5–2. The moving element of the switch is called a *pole.* There is a single make-break contact (identified as A; consequently, the switch is referred to as a single-pole, single-throw type, abbreviated as SPST. For in-home use, the switch is completely enclosed.

Known as a pictorial diagram this circuit is a two-wire type. The two wires are identified by different colored insulation. The hot lead is black; the return or neutral lead is white.

Note that the white wire is continuous. Unlike the black wire, it has no breaks for the insertion of components. The black wire consists of two sections to permit the inclusion of the switch. Current flows when the switch is in its down or *make* position. When the pole of the switch is moved up so that contact is broken

[3]*(NEC: Article 380–6; page 321)*

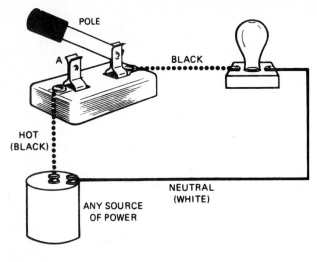

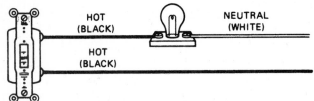

Figure 5–2. Two types of pictorial diagrams. Single-pole, single-throw switching control of a single lamp. (*Courtesy Hubbell Incorporated, Wiring Device Division*)

at point A, no current flows. The up position is called *break,* and so this switch has both a make and break facility.

The make and break positions can be readily seen for an exposed switch. An enclosed switch will often have the words *on* and *off* marked on the enclosing case. As in the case of exposed switches, *on* is equivalent to make and *off* is the same as break.

Switch Ratings

Like plugs and receptacles, switches are accompanied by voltage and current ratings, with the voltage in terms of volts and the current in terms of amperes. As shown in Figure 5–3, the rating of a typical switch carries its operating voltage and current on its metal bridge. (*NEC: Article 380–15; page 324*).

Switch T-rating

A T-rating is designated by the Underwriters Laboratories® for certain ac/dc switches, indicating that a switch may be safely used for turning circuits on and off that contain tungsten filament lamps, that is, incandescent lamps. The reason

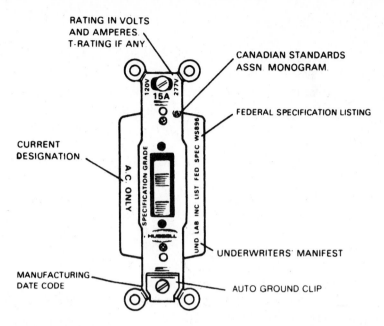

Figure 5–3. Switch ratings. (*Courtesy Hubbell Incorporated, Wiring Device Division*)

for this particular specification is that the filament of a tungsten lamp has an extremely low cold resistance. The initial current surge is about eight to ten times normal, decreasing rapidly as the filament gets hot. The heavy flow of current takes place in the first $\frac{1}{240}$ second after the switch is turned on. In this application the switch must be able to handle the surge current.

Instead of being rated for just one voltage, ac/dc switches are rated for two: 120 and 240 volts. Thus, such a switch may be used to control an incandescent lamp representing a 10-ampere load in a 120-volt circuit or 5 amperes in a 240-volt circuit. In either case the wattage rating is the same.

The greatest possibility that a tungsten bulb will burn out occurs at the 90° point in the ac cycle after the bulb is first turned on. It is at this point that the current is almost maximum and the applied voltage is close to its peak. At this time the power supplied to the lamp is about at its peak. This is a momentary condition since the current, and hence the power, will be reduced as the filament becomes hot.

The same current that flows through an electric light bulb also flows through the switch that controls it. Consequently, instead of being rated for only one amperage, ac/dc switches are rated for two, depending on the operating voltage of the controlled lamp. Thus, a switch may be rated for controlling a tungsten load of 10 amperes in a 120-volt circuit or a current of 5 amperes in a 240-volt circuit.

Specification Grade

This designation for a switch means the manufacturer has classified it as the highest-quality switch he produces. Because there is no unanimously accepted definition of the term, it does not indicate that the switch is equal in quality to any other manufacturer's specification-grade product. Other terms commonly used but not specifically defined are residential grade and intermediate grade.

Horsepower Designations for AC Switches

General-use ac switches cannot be horsepower rated. They can, however, be used to control motor loads of up to 80% of the ampere rating of the switch. An ac switch rated for 20 amperes can be used for a motor load of 16 amperes.

Double-pole, single-throw switch.[4]   Abbreviated DPST, this switch is essentially a pair of SPST types mounted on a common base and operated by the same mechanical movement, but electrically independent of each other. Its name is derived from the fact that it has a pair of blades (poles), with both moving simultaneously in the same direction to produce a make or break action.

Figure 5–4 is a diagram of this switch with the switch so connected that both lines, neutral and hot (white and black), are opened or closed at the same time. This is not good electrical practice and is illustrated in this manner simply to show the action of this switch. The switch can be used to make and break a pair of hot leads, with each possibly carrying different currents to a pair of dissimilar loads where these loads must be activated at the same time. The switch shown here is an open type, but a more common approach is one that is completely enclosed except for connecting screws.

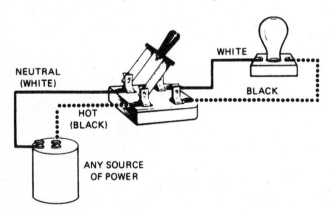

Figure 5–4.   Double-pole, single-throw switch. (*Courtesy Hubbell Incorporated, Wiring Device Division*)

[4]*(NEC: Article 380–6b; page 321)*

Single-pole, double-throw switch. This switch, as shown in Figure 5–5a, consists of a single blade mounted so that there are two independent make and break positions. As indicated in the drawing, the switch is used to operate two loads. The loads are independent of each other, but with this arrangement they cannot work simultaneously. Both loads use the same voltage source, but the load currents can be different. The amount of load current depends on which is being used.

The two loads can be operated independently or simultaneously by using a pair of SPDT switches. With such an arrangement, the total current at any time will depend on the loading: load 1, or load 2, or load 1 plus load 2.

Double-pole, double-throw switch. The switch Figure 5–5b performs the same work as that shown in Figure 5–5a. It can operate load 1 or load 2, but not simultaneously. The circuit could be simplified by connecting the white lead to each load and having just the black lead controlled by the switch. The fact that an electric circuit can be drawn and will work does not always mean it is the best circuit or the most economical.

In-line switch. Often, electrical components may not be supplied with an on-off switch for devices such as lamps of all types, illuminated Christmas decorations, lighting fixtures, signs, displays, vanity mirrors, slide projectors, electrical toys, small fans, and other low-current devices.

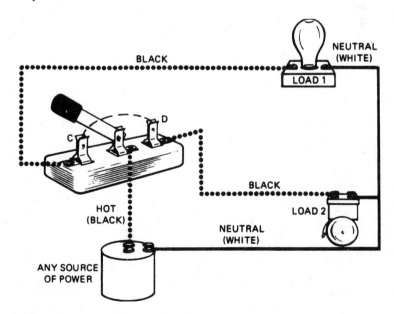

Figure 5–5 (a) SPDT switch. (*Courtesy Hubbell Incorporated, Wiring Device Division*)

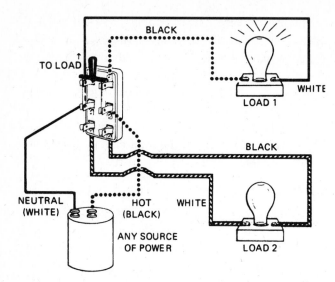

Figure 5–5 (b)   Double-pole, double-throw switching for controlling two loads. (*Courtesy Hubbell Incorporated, Wiring Device Division*)

The in-line or line cord single-pole, single-throw switch can be used to control on-off current flow through a zip-cord type of wire. To install the switch, start by cutting the connecting cord at some convenient place along the length of the wire as indicated in Figure 5–6. The position should be at a point where the switch will be readily accessible.

Slit the cord using a knife blade, but cut only one of the wires. If at all possible, cut the wire that ultimately connects to the hot lead of the receptacle.

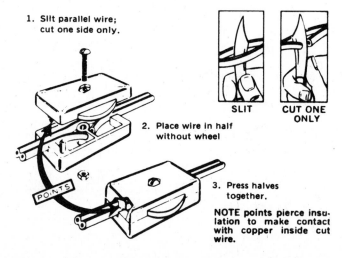

Figure 5–6.   Installation of an in-line switch. The unit is a single-pole, single-throw type. (*Courtesy Gilbert Manufacturing Co., Inc.*)

The in-line switch has a center machine screw. Remove this screw and the switch will come apart, forming two sections. Position the wire inside the switch. It will not be necessary to strip the wire since the switch comes equipped with sharp points. When the switch is assembled, these points will pierce the insulation, making contact with the cut wire.

Lamp switches. Lamp sockets can come equipped with any one of three types of switches: push switch, knob, and chain pull. They all work in essentially the same way, as SPST units. The selection is a matter of personal choice.

Toggle switch. A toggle switch is operated by moving its handle, referred to as a toggle, up or down. Most toggle types are maintained-contact units. This means the handle will remain in its position when moved up or down.

Toggle switches (Figure 5–7) are available in a variety of types such as illuminated (the handle lights when the load is turned off), pilot light (handle lights when a load is turned on), and grounded types in which a ground lug is provided on the switch to ground metal wall plates when required.

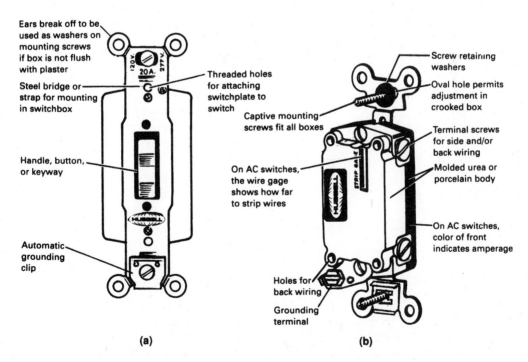

Figure 5–7. (a) Front view and (b) rear view of a toggle switch. (*Courtesy Hubbell Incorporated, Wiring Device Division*)

Lock-type switch.   This switch is operated using a removable metal key that fits into a keyway in the face of the switch. The purpose is to prevent unauthorized persons from turning the switch off and on. This is the type that would ordinarily be used in schools, hospitals, and prisons.

Momentary contact switches.   A momentary contact switch stays on or off only as long as its toggle is held by hand in the wanted operating position. A doorbell push button is an example of the momentary contact principle.

While a momentary contact position can be a single-pole, single-throw type, a single-pole, double-throw momentary contact is also available and is used to control two ac circuits. They interrupt only one wire of each circuit and are sometimes called three-position, two-circuit switches. The handle remains at its center off position until it is pushed and held up to close circuit 1 or held down to close circuit 2.

Protective switches.   Although not generally considered switches, fuses and circuit breakers are sometimes regarded as such. A fuse is a one-way switch and can only go from its on to its off condition and then must be replaced. A circuit breaker, after going into its off condition, can be reset. Circuit breakers as switches are covered in NEC 380–11. The general guidelines for switch installation are described in NEC Articles 380–1 to 380–18. These articles cover topics such as the switch enclosure, wet locations, time switches, flashers, and similar devices.

Press-action switches.   These switches are intended for ac use only and have an unusual design for the toggle, which consists of a sloping, wedge-shaped button. All it requires is a light push. This action drives a cam that makes or breaks the switch contacts.

Three-way switches.[5]   A pair of three-way switches (actually the equivalent of two SPDT) are used in Figure 5–8 to control a load from two locations. Electrical power is brought into this circuit from the two wires at the lower left, which are connected to a power source.

Four-way switches.   The circuit in Figure 5–9 is used for controlling a load from two or more locations. It uses a pair of three-way switches and one four-way switch.

Two problems occur with three- and four-way switching. The first is where to locate the switches and the second is how to run a cable between the switches.

---

[5]*(NEC: Article 380–2(a); page 320)*

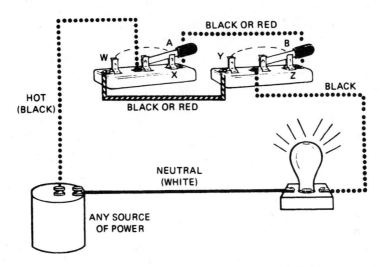

Figure 5–8. Three-way switching. (*Courtesy Hubbell Incorporated, Wiring Device Division*)

It will mean installing three electric boxes with these boxes positioned where they can be easily reached. The wiring will be supplied by a three-wire cable consisting of a black lead, a neutral white lead, and a ground wire, either bare or color coded green. The wiring can be in wall, and this will require some planning so as to avoid studs or flooring if the wiring is to go between floors. An installation of this kind is best done when a house is being constructed. Alternatively, surface wiring can be used.

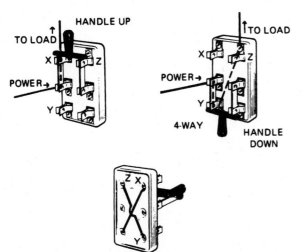

Figure 5–9. Four-way switching. (*Courtesy Hubbell Incorporated, Wiring Device Division*)

Identifying Toggle Switches

SPST toggle switch.   Although toggle switches have considerable similarity, the screws (Figure 5–10) on the sides of the switch tell a story. Of all
switches, the SPST is the most commonly used and is equipped with a pair of machine screws on the right side when facing the switch. Both of these screws have a
brass color and each is intended for the black lead of a branch line. With just this information it is possible to recognize the switch for what it is. There are a number of
other SPST switches, but this one is intended for use in a wall-type electric box.
Recognition is essential since other switches have a close resemblance.

Three-way toggle switch.   This switch (Figure 5–11) is easily recognized since it is equipped with three machine screws when facing the front. Two
of these screws are on the right side and one on the left. The function of this
switch is to control a component, usually a light, from two locations. It can also
be used for other purposes, such as making a receptacle active. The screws that
are used on the right are both brass colored. That on the left is black or may have
a copper color. Unlike the SPST, is does not have on-off markings on its toggle.

Four-way toggle switch.   The four-way toggle switch has four machine screws, with two on each side. All four have a brass color. The switch consists of a combination of three-way switches. It is shown in Figure 5–12 and, like
the preceding switch, its toggle is not marked on or off.

Double-pole switch.   This switch, shown in Figure 5–13, is sometimes
used to control high-current appliances, such as a washer or dryer, and is sometimes used to control 240-volt types. It can be confused with four-way switches

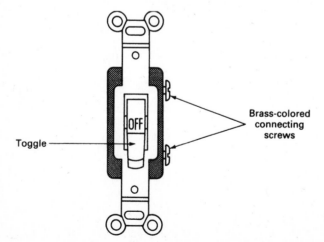

Toggle

Brass-colored
connecting
screws

Figure 5–10.   SPST toggle
switch.

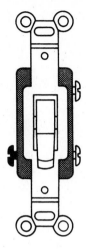

Figure 5–11. Three-way toggle switch.

since its four screws are also brass colored. The difference is that the double-pole switch has its toggle marked on and off.

Using Push-in Terminals on Receptacles and Switches

If the incoming power line from a branch enters an electric box from the rear, it may be easier to use the push-in terminals on receptacles and switches (Figure 5–14) instead of the screw terminals located on the sides. Not all receptacles and switches are equipped with push-in terminals, but when making a new installation or a replacement, it could be helpful to consider getting those that are so equipped.

There are several advantages. The connection is easier, it makes more room

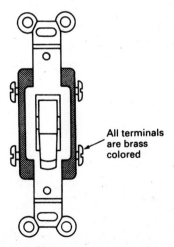

All terminals are brass colored

Figure 5–12. Four-way switch.

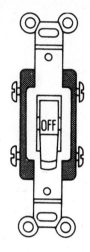

Figure 5–13.　Double-pole switch.

available inside the electric box, and it reduces the possibility that side screws may short against the sides of a metal box. However, a push-in terminal connection is not as secure as one using side screws.

To use a push-in terminal on the rear of a receptacle, strip the white and black leads about $\frac{1}{2}$ inch. Make sure all of the insulation from the stripped end has been removed. Scrape the stripped end with a penknife if necessary. Push the bare wire into the push-in hole as far as possible so that no wire is exposed. Be sure to allow some slack in the wires being connected to permit removal of the receptacle should that become necessary.

Connections to side wire screws or push-in terminals of receptacles or switches are intended for use with solid wire only. This is no problem with in-wall

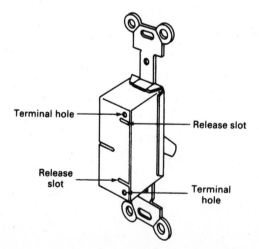

Terminal hole

Release slot

Release slot

Terminal hole

Figure 5–14.　Back holes for use as wire connections.

wiring since that is always solid conductor, but it is a consideration if surface-wall wiring is used. Such wiring should always be solid conductor as well.

Push-in terminals may be marked so as to ensure the insertion of the correct wires. If not, then those that are closest to the brass terminals are for the hot leads; those closest to the silver-colored terminals are for the neutral leads.

For switches only, two push-in terminals for the black leads are required. For receptacles, there will be four such terminals.

Immediately adjacent to the push-in terminal is a release slot. Using a small screwdriver or solid wire, push into the release slot to free a wire in a push-in terminal. Do this if the conductor in a push-in terminal is exposed.

Wiring Position[6]

Side wiring presents a problem when using metal electric boxes. The screws and their connected wires are very close to the inner walls of the box, and if the switch or receptacle is mounted off center, there is the possibility of a short. Some electricians wrap the switches or receptacles with one or more layers of plastic tape as a precaution. Using rear wiring slots helps minimize a shorting possibility. Some switches are equipped with screws for side wiring as well as rear holes.

There are some other connection techniques, but they are not as widely used as side or rear wiring.

End-wired switch. With this arrangement the connecting screws are mounted on the rear top and bottom. See Figure 5–15.

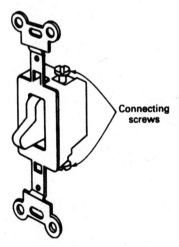

Connecting screws

Figure 5–15. End-wired switch.

[6](NEC: Article 380–6c; page 322)

Front-wired switch.   The difference here, as shown in Figure 5–16, is that the screws are mounted forward, top and bottom. This makes it possible to remove the wire connections without the need for taking the switch out of its box.

Automatic Switches

Motion switching systems.   A motion switching system is a wall switch that can be used to replace a standard toggle-type switch with a motion-sensitive infrared device that automatically turns lights on when someone enters a room and turns lights off when the room is unoccupied. This results in substantial savings in electrical energy because of automatic lights-out in empty rooms. It is estimated that lights burning in unused areas can add up to more than 60% of the total lighting cost. Also, automatic lighting means it is no longer necessary to walk into a darkened room. However, automatic lighting turn-off does take 12 minutes.

The unit operates on either 120- or 240-volt circuitry, does not require special adjustment, and mounts easily in a standard switchbox. A typical retrofit takes about 10 minutes and replaces a two-wire switch. The maximum load for the switch is 800 watts for 120 volts ac or 1200 watts for 240 volts ac.

Security light controls.   Automatic security systems can be light activated or operated by sound. Some are intended for outdoors only and use a pair of floodlights rated at 150 watts. After sunset, any movement within a previously designated area will turn the floods on. The unit is equipped with a time switch that controls the on time of the lights, which can be 1, 3, or 5 minutes. The system is equipped with a switch inserted in the hot lead, which is used to defeat operation of the lights.

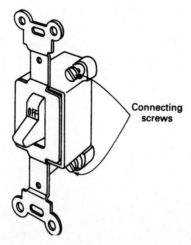

Connecting screws

Figure 5–16.   Front-wired switch.

Another security light contains a photocell that automatically turns on at dusk and turns off at dawn. The device is equipped with a pair of prongs for insertion into a receptacle. The unit can be used in a child's bedroom, in a hallway, or in a bathroom. The light will not turn on automatically when a person walks into the area nor is it affected by sound. However, it will turn on if someone passes directly in front of the unit.

Sound-activated lamp switch. Another type of switch is sound activated so that sounds turn the light on automatically. The unit is equipped with an adjustable sensitivity control and an automatic timing circuit that varies the switch's on time. No special wiring is needed, and plug in can be had with any 120 v electrical receptacle.

Sound-activated wall switch. This unit, unlike those that have been previously described, requires installation. It can be used to replace either a one- or a two-way switch. It supplies both automatic and manual operation and has adjustable fixed or automatic time control. This component has a sensitivity control setting with a light-emitting diode (LED) indicator guide.

ELECTRIC BOXES[7]

An electric box is a housing for components such as receptacles or switches. It is also used for wires that are connected, but without including components. An electric box can be plastic or metal. Metal types have knockouts to permit the entry or exit of wires or for connecting conduit to housing wires. Boxes are made for in-wall or wall-surface mounting. Those intended for outdoor use must be waterproof. Whether indoors or outdoors all boxes, regardless of application, must be equipped with a cover.

Box Size[8]

The gauge of a wire and the number of wires that may be put into an electric box are determined by guidelines established by the National Electric Code. They are governed by the size of the switch or receptacle, by the type of box, and by its shape.

[7]*(NEC: Article 370–71; page 312)*
[8]*(NEC: Table 373–6(a); page 314)*

Selecting the box.   Certain guidelines will help in counting the number of wires.

1. A grounding wire that enters a box but does not exit from that box counts as 1.
2. Leads from a fixture that enter the box and are connected to wires in that box do not add to the total.
3. Sometimes more than a single grounding wire will enter the box, but these count only as 1.
4. A number of accessories, such as hickeys, cable clamps, or studs may be used with a box but their sum is only equal to 1.
5. Each receptacle or switch counts as 1. A combined switch and receptacle count as 1.

Types of Electric Boxes[9]

Boxes are available in a number of sizes and shapes, including round, square, rectangular, and octagonal. Some are single types, while others can be ganged. Some are simply made to contain electrical connections, while others are intended to house switches or receptacles. Boxes can be metal or plastic.

Junction boxes.   The purpose of a junction box is to supply a housing for a joint or splice. Do the joint or splice after the wires have been brought into the box through an opening called a *knockout*. After joining the wires in a tap or splice, and, if possible, using a wire nut (preferably with an outer layer of electrician's tape), cover the box with a blank plate. Aside from openings for machine screws to hold the plate in position and the openings for incoming and outgoing wires, there should be no holes in the box or its cover. Any unused knockouts should have a cover (Figure 5–17). The junction box should be enclosed on all four sides. For future work it would be helpful to letter a note to be pasted on the outside cover of the box supplying some data about the internal connections.

Figure 5–17.   Cover for knockout of metal-type electric box.

[9](*NEC: Article 370–1; page 302*)

Figure 5–18. Octagon box.

Octagon boxes.   The shape of an electric box (Figure 5–18) can help decide its ultimate use and its size can help determine the number of wires it may contain. Octagon boxes are usually made of steel and can have $\frac{1}{2}$inch diameter knockouts. They can also be equipped with cable clamps built into the box and used to hold one or a pair of cables firmly in position by turning a machine screw.

Rectangular boxes.   These are designed for switches and receptacles and are available with or without built-in cable clamps (Figure 5–19). It is possible to gang such boxes by arranging them to work side by side. This is done by removing a side plate from each box, thus doubling the available volume (Figure 5–20).

Beveled corner boxes.   If possible, mount junction boxes where they will always be easily accessible. Do not plaster over them or conceal them behind walls. The advantage here is that the junction box is a natural takeoff point for a new branch should it become necessary to expand the wiring setup. The beveled corner box (Figure 5–21) is so called since its rear section is beveled.

Side-bracket boxes.   This is a square or rectangular type and is so called since it has a bracket on one side for mounting the box. The box can be used for switches and receptacles and has either $\frac{1}{2}$ or $\frac{3}{4}$-inch knockouts. It can be used with Romex, BX, or conduit.

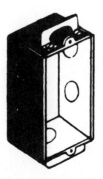

Figure 5–19.   Rectangular box.

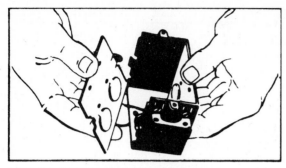

Figure 5–20. Ganging rectangular boxes.

The side bracket can be to the left (Figure 5–22) or the right of the box. These boxes are generally used only in new installations, since it is difficult to install the brackets once the walls are completed.

The advantage of side brackets is that they permit mounting of the box so that it is always at the correct height and depth. Side brackets are also convenient for mounting boxes on wooden crossbeams in an attic or basement. Since these areas are exposed, the side brackets under these circumstances can be used in either new or old homes.

Round boxes.   While round boxes (Figure 5–23) may be available in electrical supply stores, they are not commonly used in home wiring. Because of the curvature of the box, it is difficult to fasten BX or conduit to the box, although Romex can be held in place using a suitable clamp. Round boxes are sometimes

Figure 5–21.   Beveled corner box.

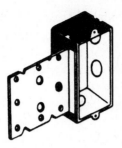

Figure 5–22. Box with side bracket.

used to install ceiling lamps. They require a round cover plate made specifically for such boxes.

Through boxes. Some boxes are through types and are used for installing switches or receptacles on both sides of a wall in back-to-back fashion. Unlike ordinary boxes, these are not completely enclosed but are open both front and back. The advantage of using a through box is that it is both convenient and economical. Installing such a box can be a time saver.

Electric boxes for outdoor use.[10] Use weatherproof boxes for outdoor installation. The distinguishing characteristic of these boxes is that they are equipped with weatherproof snap covers. The covers are spring loaded and must be lifted to permit access to the receptacle housed in the box. When the plug is removed, the cover snaps shut, protecting the receptacle against the weather.

Shallow box. In some instances the space available behind a wall may not be enough to accommodate an electric box. Some boxes are made whose depth is much narrower than usual, but the volume of such boxes permits the use of only one switch or receptacle with just one branch line cable.

Ceiling boxes. All boxes, whether in-wall or on-wall (or ceiling) types, face the problem of mounting. The problem is somewhat limited for those boxes that can be mounted in wall against a stud. An on-wall box is not too attractive

Figure 5–23. Round box.

[10](*NEC: Article 370–15; page 303*)

and must supply a channel that will cover the wires and supply some protection as well. A ceiling box can be mounted against a wooden ceiling beam with the dry wall cut away to permit access to the box. If the ceiling is made of cement, the box must be held in place with lead anchors, which require drilling. For a plaster ceiling, plastic anchors can be used.

For ceilings adjustable bar hangers can sometimes be used, depending on the construction above the ceiling proper. The bar hanger supplies support for the box. Use a bar hanger that is adjustable so that it can be moved any small distance between the joists (Figure 5–24).

Do not use conduit, BX, Romex, or any other kind of wire as a box support. All boxes must be securely supported independently of any other branch circuit accessories.

Box Materials

Boxes are made of two materials, metal such as steel (possibly galvanized) and plastic. Polyvinyl chloride (PVC) and fiber glass are used for the electrically non-conductive boxes, and these are only used with plastic conduit. As in the case of metal boxes, the plastic types are equipped with thin circular sections that can be broken away using a screwdriver and a hammer, thus supplying a knockout. Plastic boxes can be fastened to a stud by using nails. Since the plastic box does not conduct, it is necessary to have a continuous ground wire as well as a continuous neutral.

In the event of a fire, a plastic box does not supply as much protection as a metal one. The plastic will deform or may simply drop away, exposing wires to flame. NM cable is the only kind that may be used with plastic boxes.

In some municipalities, plastic boxes are forbidden. Plastic boxes can be purchased in ganged form, but it is not possible to disassemble them. (*NEC: Article 370–40, page 311*).

Plastic electric boxes are mounted in the same manner as metallic types.

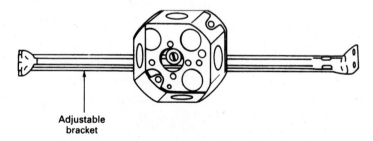

**Adjustable
bracket**

Figure 5–24.   Adjustable ceiling box bracket.

More care is required for plastic boxes since they can suffer more damage. Non-metallic boxes are lighter and can be kept in position by spring clips positioned along the sides. This method is useful for adding a box to an existing home and nailing or screwing a box to a stud.

Box Extenders

There are two ways of increasing the volume of a box. The first is to use a pair of identical boxes whose sides can be removed. When these boxes are joined, the original box volume is effectively doubled. Another is to use a box extender. This resembles a box with its back removed. It is put up against the original box and held in place by machine screws through the flanges of the boxes.

ELECTRIC BOX COVER PLATES[11]

When a switch is set up in a possibly moist environment, a weatherproof cover plate and weatherproof enclosure for the switch are recommended. With this arrangement, no part of the switch is exposed to the elements. The plate is made of a material that can be depressed in the area of the switch so that it can be activated without exposure. Instead of an opening in the cover plate, there is a resilient bubble in the plate just over the switch button. To activate the switch, all that is necessary is to press the bubble.

Another switch waterproofing technique consists of a cover plate with a swing-type cover. A coil spring around the cover hinge holds the self-restoring cover in a closed position. A waterproof gasket prevents moisture from seeping between the plate and the box.

Types of Cover Plates

Toggle-switch cover plate.   This plate (Figure 5–25) has a cutout for the toggle and two holes to accommodate $\frac{6}{32} \times \frac{1}{4}$ inch machine screws for holding the plate in position. When properly in place, the plate will be flat against the wall with the toggle extended its maximum distance outward. If the electric box is too deeply recessed, the toggle will be out a shorter distance and may be difficult to operate. With the retaining screws at their maximum in position, the plate should

[11](*NEC: Article 370–25a; page 309; Article 410–56d; page 370*)

Figure 5–25. Toggle switch and its cover plate.

be firm but have no bulges. If the plate is a plastic type, excessive force on these screws can cause the plate to crack.

Receptacle cover plate.[12]   Also known as face plates, cover plates for duplex receptacles (Figure 5–26) come equipped with a single, center-positioned hole to accommodate a $\frac{6}{32} \times \frac{1}{4}$ inch machine screw for fastening the plate to its receptacle, thus holding the plate in position. While there is just a single, centered holding screw, the outer edges of the receptacle should fit snugly enough in the plate to supply some additional holding strength.

Clock cover plate.   Electric clocks are accompanied by their own length of line cord plus a plug. The receptacle (Figure 5–27) for these clocks is recessed to hold both the plug and wire. This permits the clock to hang flush with any wall surface.

Miscellaneous electric box covers.   As is usually the case with components for electrical work, there is a large variety of cover plates available for switches, receptacles, and other electrical appliances. Figure 5–28 shows a number of these.

Figure 5–26. Duplex receptacle and its cover plate.

[12]*(NEC: Article 380–9; page 322)*

Figure 5–27. Clock cover plate. It is recessed to hold both plug and wire. This permits the electric clock to hang flush with the wall surface.

Figure 5–28. Cover plate (upper left) and types of electric box covers. (*Courtesy Mulberry Metal Products, Inc.*).

Cover plates for weatherproof receptacles and switches.   Figure 5–29a shows the cover plate for a single outdoor receptacle, but plates for duplex types are also available. They are made of metal and have spring-loaded circular covers. The single types are held in place by two screws; those for duplex receptacles use a single, center-mounted screw. Figure 5–29b shows the cover plate for a switch.

Cover Plate Colors and Finishes

Cover plates (also known as face plates or wall plates) are commonly supplied in colors such as white and brown and much less often in black. In some instances, though, they are now regarded as part of a room's decor, and so a variety of materials, colors, and finishes is available. These include bronze (statuary and satin), brass (satin and polished), aluminum (clear, black, dark brown anodized, and satin); and stainless steel (satin and polished); various finishes such as semigloss, matte, smooth, hard baked, wrinkle, mirror, and dull; and painted metal, such as red or black (with red preferred for emergency use), wrinkle-painted metal in white, brown, or ivory, antique copper and gold, and hammered copper.

Cover Plate Sizes

Typically, cover plates are supplied in three sizes referred to as standard, maxi and jumbo. Figure 5–30 shows the most commonly used sizes for toggle switches. For a switch the cover plate uses two mounting holes; for a standard duplex receptacle it uses one. There are some variations though and for quadruple-type receptacles the plate has four holes and also has four for duplex switches.

The position of through holes for boxes housing switches can vary as shown in Figure 5–31. Most commonly they are 3. 281 inches apart, but there are others as indicated in the illustration.

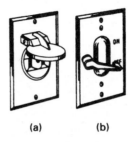

(a)   (b)

Figure 5–29.   (a) Cover plate for outdoor receptacle; (b) cover plate for outdoor switch.

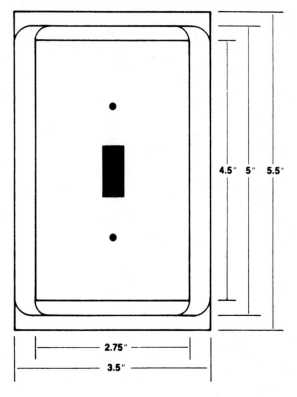

Figure 5–30. Dimensions for commonly used cover plates. (*Courtesy Mulberry Metal Products, Inc.*)

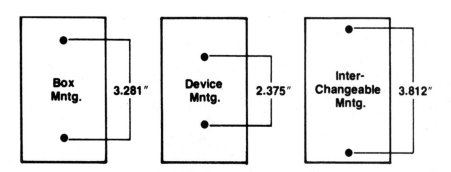

Figure 5–31. Location of screw mounting holes for switch cover plates. (*Courtesy Mulberry Metal Products, Inc.*)

Plates for Special Situations

Narrow plates.   A typical narrow wall plate is shown in Figure 5–32, but it can be narrower and shorter as well. These plates find application in office partitions, trailers, mobile homes, and marine installation. Available heights are $4\frac{1}{2}$, $6\frac{1}{8}$, and $8\frac{1}{8}$ inches. To avoid getting a plate that cannot be mounted, measure the location of the holes on the switch or receptacle or on the old cover plate.

Jumbo cover plates.   If the wall behind the old plate and its surrounding area has been damaged, it may be possible to cover it by using a jumbo cover plate. Just make sure that its mounting hole or holes will match those on the receptacle or switch. You may want to match the color or finish of other plates in the same room or other rooms.

Deep plates.   In some cases the position of the electric box will determine the choice of plates. It is not always possible to mount an electric box in the most desirable location, for example, if the installation of the box means it will extend from the wall. This may be just a small amount, but it can be surprisingly noticeable. In that case, select a deep plate, such as those shown in Figure 5–33. The depth can be (a) $\frac{9}{16}$ inch or (b) $\frac{3}{8}$ inch. They are available for one-, two-, or three-toggle switches.

Emergency cover plates.   In some places, such as hospitals, it is sometimes essential to have a receptacle or switch that is to be used only in the event of an emergency. Known as a dedicated line, it is to be used only for that purpose. To emphasize its function, its cover plate is identified by the word "Emergency" printed across the top. It is often identified in black, and since most plates do not

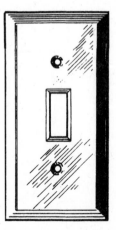

Figure 5–32.   Narrow cover plate. It is available in these heights: $4\frac{1}{2}$, $6\frac{1}{8}$; and $8\frac{1}{8}$ inches. (*Courtesy Mulberry Metal Products, Inc.*)

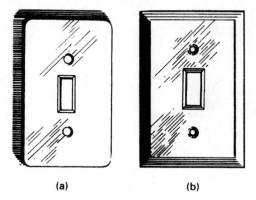

(a)   (b)   **Figure 5–33.** Deep plates.

use this color, it emphasizes its emergency message. Some emergency plates use white, but this color is not as striking since ordinary plates also use white. Emergency plates are designed for single-toggle switches and duplex receptacles (Figure 5–34) as well as for double duplex receptacles.

Plates for oil and gas burners. Single-toggle switches are often mounted in an electric box on metal conduit with the conduit fastened to the side of a gas or electric boiler and containing a power line that leads to the boiler. The plates, indicated in Figure 5–35 are easily identified and are colored in red. Suitably marked plates are also used for large air-conditioning systems.

To avoid accidental use of these switches and as a reminder that they are restricted, the face may be covered with a guard, as in Figure 5–36.

Mounting the Cover Plate

All plates are held to their associated switch or receptacle by machine screws. Figure 5–37 shows the assembly for toggle switches. The plate is mounted after the switches or receptacles have been wired, tested, and then positioned in their box. The only purpose of the screws is to hold the plate securely in position. It is not intended to support, in any way, the component inside the electric box.

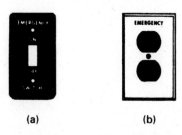

(a)   (b)

Figure 5–34. Emergency cover plates. (a) Toggle switch plate; (b) duplex receptacle plate. (*Courtesy Mulberry Metal Products, Inc.*)

Figure 5–35. Oil burner cover plates. The plate at the left is intended for a round electric box. (*Courtesy Mulberry Metal Products, Inc.*)

If there are a number of components inside the box, such as the three toggle switches in Figure 5–37, a number of mounting screws will be required, one for each receptacle and a pair for each switch. Do not omit any screws.

The screws holding a cover plate in position should fit smoothly and easily into their respective tapped holes on the component inside the electric box. These components have some free play by adjusting the screws that hold them in position in the box. It may be necessary to loosen and readjust them if the plate-holding screws do not fit properly.

SOCKETS[13]

A socket is a device for holding a lamp, a screw-in type single receptacle, or a fuse. The inner wall of the socket may be smooth or threaded. Some socket terminals are connected to a source of line voltage and so supply ac to the device

Figure 5–36. Emergency switch guard. (*Courtesy Mulberry Products, Inc.*)

[13](*NEC: Article 410–3; page 358*)

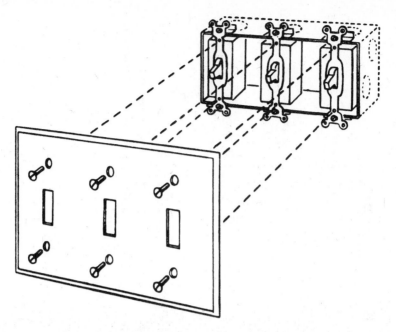

Figure 5–37. Three-gang switch plate on a flush-mounted three-gang switchbox. (*Courtesy Hubbell Incorporated, Wiring Device Division*)

inserted in the socket. Some sockets have exposed terminals and, depending on the amount of source voltage, can be a shock hazard. Sockets are available in a number of sizes, ranging from small ones used for flashlight bulbs to large ones for accommodating electric light bulbs. Some sockets are single types, such as those in flashlights or lamps, and some are double types, as in fluorescent fixtures.

Sockets are listed in the National Electrical Code under the generalized heading of Fittings (Article 100, page 29) where the device performs a mechanical rather than an electrical function. The purpose of a socket is to allow the easy connection of some electrical device to a voltage source, either ac or dc, but without using any part of that current for its proper functioning.

Types of Sockets

Cleat socket. Figure 5–38 is a drawing of a cleat socket. Sockets of this type are wired across an ac power line, that is, in parallel with it. The two wires, a black lead and a white lead, are stripped and then fastened under the screws of the socket. The socket has no accommodation for a ground lead.

Exposed cleat sockets were widely used at one time when ac wiring con-

Figure 5–38. Cleat socket. Exposed terminals for connecting ac line is possible shock hazard.

sisted of a pair of on-wall wires. Although exposed cleat sockets are still sold, in various areas they are a violation of local electrical codes. Their advantage is that they are easy to install and connect. Their disadvantage is that they are not switch equipped, and so the lamp or other electrical device must be removed from the cleat socket to turn it off. Figure 5–39 is a cross-sectional view of this type of socket. When a lamp is screwed into a cleat socket, the threaded portion of its base makes contact with the similarly threaded section of the cleat. The other contact is via a metal bit on the central part of the lamp base. These two contact areas are insulated from each other.

Base-pull socket.   This socket is an improved version of the cleat. Electrical power is delivered to connections beneath the base, and so a possible shock hazard is eliminated. Another improvement is that it uses a pull chain-operated switch, and so electrical power is easily turned on and off.

Figure 5–40 illustrates one socket of this type. The base of the socket is equipped with through holes, and so the unit can be mounted on an electric box. If the pull chain is too short, it can be extended by an add-on length of chain or a string. Pull-chain sockets are often used for basement installations.

Weatherproof sockets.[14]   The weatherproof socket shown in Figure 5–41 is intended for temporary outdoor use. Unlike cleat types, which are constructed of porcelain, the weatherproof socket is made of rubber or a rubber substitute. The socket is equipped with a pair of leads for connection to wires leading to a

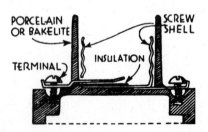

Figure 5–39.   Cross-sectional view of cleat socket.

[14](NEC: Article 410–4a; page 358)

Figure 5–40.   Base pull socket.

power source. The unit has only two wires, is not grounded, and is not recommended. It is advisable to make offset connections, with one of these about an inch or two away from the other so as to minimize the possibility of an accidental short. Make sure each connection is completely covered with a wire nut and electrical tape.

This arrangement is for temporary use only. Its purpose is to supply electrical power outdoors in the event an outdoor electrical receptacle is not available. After the socket has served its purpose, disconnect it from the power source and bring the socket indoors. Do not use it during rainy weather or when outdoor conditions are damp.

A better arrangement than this makeshift setup is to use an outdoor extension cord that is waterproof-socket equipped. These are equipped with a metal cage for an electric light bulb, not only to protect the bulb but as a guard against a burn from the hot lamp. The extension cords are available in various lengths, are marked with their wire gauge, and come equipped with a molded plug. They are intended for connection to a three-terminal receptacle.

ELECTRICAL HARDWARE[15]

Electric hardware includes small but important parts used for fastening plastic sheathed cable or BX. Although they are not electrical in the sense that they carry current, they make an important contribution to the stability of an electrical system.

Figure 5–41.   Weatherproof socket.

[15]*(NEC: Article 333–7; page 243)*

Captive screws.   There is little that is more exasperating than dropping a tiny machine screw only to have it disappear. The problem is that machine screws used in electrical work are often no more than $\frac{1}{4}$ inch long, and it takes just one or two turns of a screwdriver to remove them.

To prevent their loss, some electrical hardware comes equipped with upset screws, more popularly known as captive screws. This type of screw has a somewhat larger diameter at the end of the screw (the end opposite the screw head). When turning the screw outward, the upset end will prevent the screw from being completely removed. Enough of the screw can be turned to permit connecting or removing a wire.

The advantage is that the upset prevents possible loss of the screw. The disadvantage is that if you use a strong turning force the screw will come out anyway, and in so doing may damage the threads in the terminal to which it was attached. This means a replacement screw may not fit or may fit very loosely.

Staples

While BX and conduit may look strong, and they are, their housing is intended to protect wires and not to act as a support. Also, both BX and conduit must be supported so that there is no mechanical strain.

There are two types of cable staples, those in which the two arms of the staple are parallel to each other and those whose arms are offset. The straight or parallel-arm staple is the easier to work with. Put it into position over the cable and then hammer the flat portion of the staple until it makes firm contact with the BX. While BX has been largely replaced by conduit, it may sometimes be necessary in older homes to reinforce the BX support.

Offset staples can be handled in the same way as the parallel-arm type, but if you should later change your mind and decide to remove the staple, you will find it more difficult. It does, however, supply a secure anchorage.

Insulated staples.   Use insulated staples for low-voltage wire, but not for lamp cord or for any other wires carrying current from a 120-volt ac line. An insulated staple is smaller than cable staples and has a small bit of nonconductive material across the inside of its top, flat portion. Homeowners sometimes use these insulated staples for tacking lamp cord to baseboards, which is not a good procedure since the insulation on the staple can fall off, and it then becomes possible for the staple to cut through and short the wires. Insulated staples can be used with Bell wire.

Cable straps.   Cable straps and cable staples perform the same function of holding cables in a fixed position, but they differ in their construction and in the way they are mounted. Cable straps are available as nonmetallic or metallic types. The advantage of a nonmetallic cable strap is that it will not cut through and short wires. Its disadvantage is that it must be mounted with screws. To save time and effort, however, some straps have one-screw mounts.

A two-screw strap is sometimes called a full strap, while straps using a single screw are termed half-straps. Either the full or half-strap can be used for conduit. The full-strap type supplies stronger and firmer support, but sometimes there is no choice but to use the half-strap. There may not be enough room for a full strap, or the conduit may be in such a location that it is not possible to use the full strap.

Chapter 6

Practical Wiring[1]

PART I

Electrical wiring begins with a collection of separate components: a fuse or circuit breaker box, wire for the formation of branch circuits, electrical boxes, switches, receptacles, and miscellaneous hardware. These all start as isolated items but, when correctly joined, they supply electrical energy, controlled energy that makes possible heating, air conditioning, lighting, electric washing machines, electric dryers, toasters, broilers, electric stoves, radio, high fidelity, television, and much more.

SHUTTING OFF POWER AT THE SERVICE BOX

Some basements have damp floors following a heavy rain or may be customarily damp because of water seepage through their walls. If it is necessary to close the main power breaker or fuse, dry the floor near the service box as much as possible and then put down a dry scrap rug, a wooden panel, or boards. Wear gloves or

---

[1](*NEC®: Article 300–6c, page 166*)

172

use a thick dry towel to open the door to the box. Use one hand only and keep the other behind your back. Open the main circuit breaker. All lights and appliances should then be off.

If the service box uses a fuse block equipped with a handle, pull on this to shut off power. The box may be equipped with several fuse blocks; if this is the case, remove all of them. An older type of box may have an outside metal handle with the handle in the up position. Pull down on the handle to shut the power off before doing any electrical work.

WIRING PROBLEMS

The location of the service box, the installation of the various branch lines, and the placement of receptacles is the work of a licensed electrical contractor at some time during the construction of a building, whether it is a home, an office, or a farm building. It is rare for the ultimate occupants of a home to be consulted, and so the users of the premises must adapt to an existing condition. But even if the wiring plans are made available, they are subject to changes since electrical component usage is always increasing.

Aside from wiring faults, problems can exist whether the home or building is new or not.

1. *Physical placement of receptacles.* These are not always where they should be. Installing new ones can involve breaking into existing walls and snaking wires behind them. Or it can mean making an on-wall arrangement.

2. *Selection of a method for increasing the ampacity of an existing branch line, (NEC: Table 400–5B; page 346).* The occupants of a new home or an old one seldom pay much attention to wiring requirements. The attitude is generally that all wires are alike and that they can carry any amount of current. The subject of wire gauge is a mystery. The need for installing a new branch line is often a surprise. *(NEC; Article 110–6; page 39)*

3. *Need to install more receptacles to accommodate more plugs.* Additional receptacles may be required to permit the use of more appliances. The inclusion of new electrical devices such as motor speed controls, light dimmers, ground-fault devices and/or surge suppressors is now part of the modern electrical setup. These may be in addition to a home entertainment system consisting of a large TV receiver and a high-fidelity system, plus electric warning systems and more powerful home appliances. All these can involve modification of an existing wiring setup.

SHUNT AND SERIES CONNECTIONS

Electrical wiring consists of three types of arrangements: shunt (also known as parallel), series, and a combination of shunt and series. The diagram in Figure 6–1 shows the shunt method. The three lamps are in parallel with each other, with this combination connected to the ac power line. One wire of the parallel lamps, a wire color coded black, is connected to a single-pole, single-throw (SPST) switch. The other lead, a white wire, called the neutral lead, is connected to the other side of the lamps. Note that the white wire is continuous, unlike the wire color coded black.

When a switch is closed, an electrical current surges back and forth through both wires, the lamp, and the switch. The lamps are referred to as a load, but all can have equal or different current requirements. The switch is not a load and has no influence on the amount of current flowing in its circuit. The wattage of the lamps determines the current flow; the only function of the switch is to control that flow, to turn it on or off.

Since the voltage source is ac, so is the current flow. This means the current flows from the source, through the switches, through the lamps, and then back toward the source. When the polarity of the voltage reverses, so does the current flow. It now moves from the source, opposite to its original direction, through the lamps, through the switch, and then back to the source.

At no time during this sequence of events is there a loss of current. The significance of this is that the same amount of current flows through the hot wire and the neutral. If the neutral wire is connected to a ground wire, it will be in shunt with it. Since the neutral is in parallel with the ground wire, these two conductors will share the current, but not necessarily in equal amounts.

Circuit Variations

While the circuit in Figure 6–1 is very simple, it is possible to draw some variations. The load could be three lamps, as shown in the drawing, or just a single

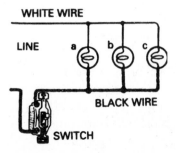

Figure 6–1. When the switch is closed the three lamps, a, b, and c, are shunted across the power line. The three lamps are turned on and off at the same time.

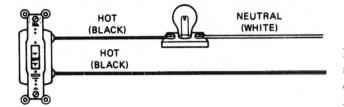

Figure 6–2. SPST control of a single load. (*Courtesy Hubbell Incorporated, Wiring Device Division*)

one, as in Figure 6–2. Note that this drawing follows two basic rules: the load is connected to a black lead and also a white lead. This applies to any load.

The switch is connected only to the black lead and is actually in series with it. However, the switch in Figure 6–3 does have three terminals: two for the black lead and one, a nickel-plated terminal, for a white lead. The nickel-plated terminal can be used as a ground connection. The black and white leads shown in a downward direction are to be plugged into a branch line, possibly using a plug inserted in a receptacle.

Another diagram along the same lines is shown in Figure 6–4. This is still the same circuit and also uses a three-terminal switch. Note that the neutral wire is not connected to the switch. The third terminal of the switch is a ground connection. How a circuit is wired, then, depends on the component that is used, and, as shown in Figures 6–2 through 6–5, they are not all alike.

Voltage across a parallel circuit. When two electrical components are wired in parallel, the same voltage will appear across each. This does not apply to the amount of current flowing through them.

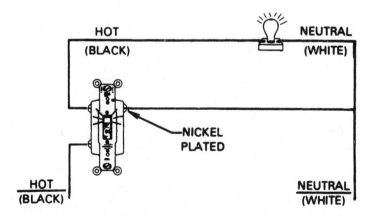

Figure 6–3. Three-terminal SPST switch for load control. (*Courtesy Hubbell Incorporated, Wiring Device Division*)

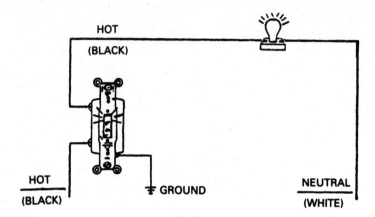

Figure 6–4. Three-terminal receptacle with one terminal grounded. (*Courtesy Hubbell Incorporated, Wiring Device Division*)

Shunt Rules

1. All devices that require current to operate are shunt connected. Electric lights and electrical devices such as toasters, heaters, electric stoves, fans, air conditioners, and broilers are current operated; that is, they must have current flow through them and hence must be shunt connected. Every shunt-connected device is a load.

2. While a current will flow through a switch, fuse, or circuit breaker, these

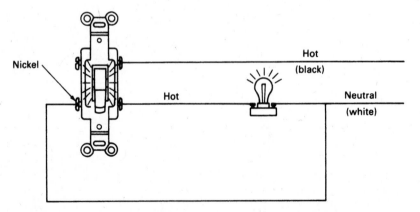

Figure 6–5. This is the most common arrangement of SPST wiring. (*Courtesy Hubbell Incorporated, Wiring Device Division*)

components do not depend on a current flow through them. They are independent of current flow. An electric bill does not have any costs charged to these series-wired components. Their only function is to control the current flowing through them. A fuse, circuit breaker, or switch connected in parallel across a power line represents a short circuit.

3. Never connect a wire or any other current-carrying conductor such as a metal tool in shunt with the power line.

4. All receptacles are wired in shunt with each other and with a branch power line. All electrical devices when plugged into a receptacle are automatically in shunt with all other electrical devices and are in shunt with the main power line and all its branches.

5. The larger the number of active appliances that are plugged into receptacles, the greater is the total load and the greater the total amount of current flowing through the connecting hot, neutral, and ground leads.

Series Rules

Never wire appliances in series (Figure 6–6). To do so means dividing the line voltage for each of them. For three identical 120-volt bulbs, each would receive one-third of the voltage and none of the bulbs would light. However, there are some rare exceptions. Thus, the tiny bulbs used for Xmas trees are sometimes wired in series. If the line voltage is 120 and 10 bulbs are used, each would be rated at 120/10 or 12 volts. The disadvantage is that if any one of the bulbs becomes defective none of the bulbs will light.

Always wire fuses, circuit breakers, and switches in series with the black lead of a power branch. While none of these components works as a load, each must be designed to handle the maximum possible current demanded by the total estimated shunt load.

All electrical circuits are series-parallel types (Figure 6–7). All loads from as little as an electric shaver to as much as a vacuum cleaner are connected in shunt. The two, series and parallel, work together.

Figure 6–6. Loads in series. An open, as in drawing at right, means current stops flowing in all loads.

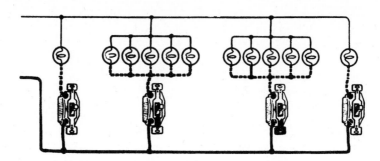

Figure 6–7. Series-parallel circuit. The switches are in series with the lamps. When the switches are closed, all the lamps are in parallel with the power line.

Ground[2]

At one time the wire color coded white was referred to it as a ground lead. It was supposed to be connected to earth, and while it was intended to be continuous, it had various amounts of resistance along its length. A water pipe was presumed to represent a good ground, but this assumption was not always valid. Water enters a home through a meter and in some instances the meter structure represents a high resistance. Bypass the meter by shunting it with a heavy cable. But that is just one trouble spot. Water pipes are made of sections, and if these sections are corroded, they can present high-resistance joints. While these joints can also be jumped by heavy gauge wire braid, not all are accessible.

The solution to this problem is the inclusion of a wire whose insulation is color coded white. Known as a neutral, it is ultimately attached to a neutral terminal in the service box. The neutral may make one or more connections to the ground wire.

Switch and Receptacle

The earlier drawings of switch connections are partial since they do not include the associated receptacles. The input to a receptacle consists of a cable that includes the black lead (hot lead) and a white lead (neutral), plus a ground wire. The receptacle is a way of breaking into a branch power line, and its purpose is to make power available to a switch or receptacle.

Figure 6–8 shows a SPST switch and its electric box connections. Except for end of the run receptacles, they all have power-input and power-output cables.

[2]*(NEC: Article 250–42E; page 132)*

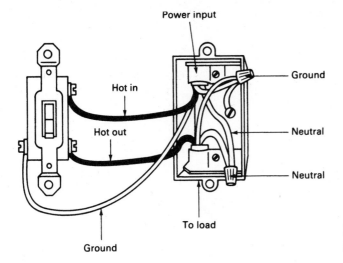

Figure 6–8. SPST switch and electric box wiring.

WIRING EXAMPLES

Connecting a Ceiling Light[3]

A drawing of a switch-controlled light can give a false impression since it does not include any idea of the mechanical work involved. Installing an electric box may mean cutting into a wall, fastening the box to a stud, and fishing wires from some other electric box. Often, the mechanical work is tedious and requires substantial time before the actual wiring can be done.

Figures 6–9a and b shows the work involved in setting up and connecting a ceiling globe lamp. The lamp is supplied prewired, but it needs to be supported from an electric box in the ceiling. The lamp may be equipped with a pull-chain on-off switch, or it may need to be connected to a SPST switch mounted in a nearby electric box. Before installing, turn the power off by removing the fuse or opening the circuit breaker.

Remove the globe and the light bulb. Insert the stud in the electric box and turn it about six to eight turns. Mount the metal strip on the stud and hold it in place with the locknut. Bring the canopy up to the electric box and hold it in place with the pair of mounting screws. Use just a few turns.

Bring the black and white wires out of the electric box. Fasten the black and white wires from the fixture to the wires from the electric box, black to black, white to white. Wrap the bare wires around each other and make a tight connection. Cover the connection with wire nuts. Push the wires into the box and tighten the mounting screws. The canopy should fit snugly against the ceiling.

[3]*(NEC: Article 410–G; page 367)*

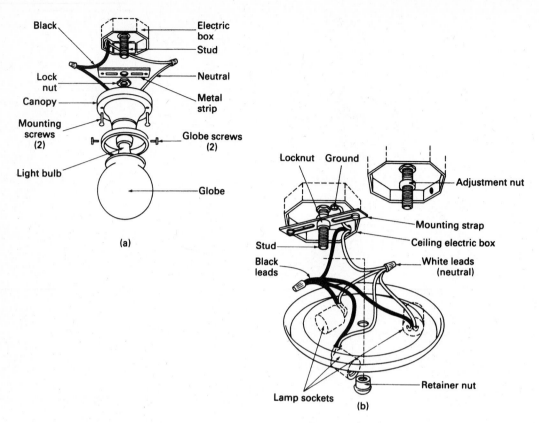

Figure 6–9.   (a) Wiring a single ceiling light. (b) Wiring three ceiling lights in the same fixture.

Insert a light bulb into the fixture socket. Restore power and turn the switch on to check the results. Replace the globe and tighten the globe screws.

Not all fixtures are alike. Some have a pull-chain switch and so there is no need to use an electric-box-mounted toggle switch. Others have more than one light, but these are prewired and the number of connecting wires remains the same.

How to Install an Attic or Basement Light

The arrangement in Figure 6–10 is similar to that shown in Figure 6–9, but it is much simpler from a mechanical viewpoint. Electrically, it is practically identical. The setup consists of an electric box receiving an input power cable. The item to be used is a mounting plate equipped with a lamp socket and an on-off pull chain switch. Shut the power at the service box. Connect the black wires from the branch

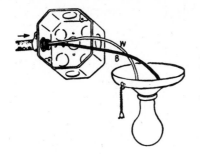

Figure 6–10. You can install a new box and get the convenience of a pull-chain fixture. The pull-chain can be equipped with a mounting plate and held to the box with just two machine screws.

and the light assembly and also the white wires from the same sources. Cover the wire ends with wire nuts. Some electricians wrap the wire nuts with plastic tape.

The pull-chain fixture will have a pair of machine screws. Use these to fasten the fixture to the electric box. Restore the power and test by using the pull chain.

Pull-chain fixtures of this kind are often made of porcelain. They may be equipped with one or two built-in receptacles that are prewired. These receptacles are not affected by the pull chain and are always ready for use.

How to Illuminate a Receptacle

In some cases a receptacle may be installed in a dark area, making plug insertion difficult. This problem can be solved in several ways. A night light can be installed, but that occupies one-half of the duplex receptacle. A flashlight can also be used, but that is not always convenient. Still another technique is to use a switch. As indicated in Figure 6–11, the receptacle is always live. The light can be turned on or off by using the switch.

How to Wire a Junction Box[4]

Wires carrying electrical power that are to be joined must be positioned in an electric box referred to as a junction box. As a first step, make sure the power is turned off by removing the fuse for the branch line or by switching the corresponding circuit breaker to its off position.

Select a suitable location for the junction box, possibly on an exposed joist or beam. The junction box may be used as a starting point for an additional power-line branch or as a possible location for some future receptacle or switch. The junction box can be either plastic or metal.

Cut the wires of the existing cable, pass them through a connector. (See Figure 6–12) and bring the cut ends into the box. Inside the box is another connector

[4](*NEC: Article 370–17; page 305*)

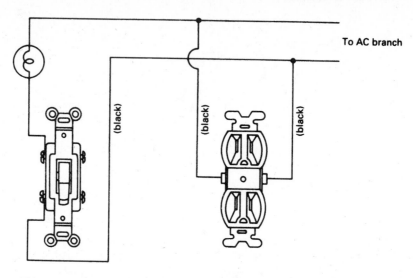

Figure 6–11. SPST switch for control of a light bulb and a duplex receptacle. A dark line in a drawing is sometimes used to indicate the hot lead. This drawing technique is used mostly in older diagrams.

which must be tightened. The cable must have enough slack to enable you to do this. Remove about ½ inch of insulation from the ends of the black leads. If any insulation clings to the exposed copper wires, clean them with a knife, sandpaper, a fine-tooth file or scrape them with diagonal cutters. Form a splice using pliers to make the splice as tight as possible.

Repeat this action with the white wire. If the cable has a ground wire, strip about ½ inch of insulation from its end and connect it to any convenient grounding

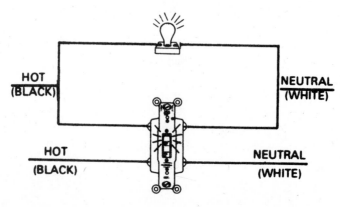

Figure 6–12. DPDT switching for controlling two loads. (*Courtesy Hubbell Incorporated, Wiring Device Division*)

terminal in the metal box. This not only grounds the box but supplies continuity for the ground wire. If the ground wire is bare, make a single turn around a grounding screw. Tighten the screw. Cover the individual white and black wire ends with a wire nut. Cover the box with a face plate.

How to Tap in on a Switching Circuit

When deciding to install an additional receptacle, it may be possible to tap in on an existing switch and load line that is conveniently located, as in Figure 6–11. The receptacle is wired directly to the branch line, and since it is in shunt with the hot lead and the neutral, it is always live. It is not affected by the setting of the switch. When the switch is closed, the lamp and the receptacle are directly in shunt.

How to Control Two Loads with One Switch

The switch used in Figure 6–12 is a DPDT. The upper two switch terminals are for connection to one load; the lower two are for a second load. The switch is a three-position type, and when in its center setting, both loads are turned off. The two loads can be controlled sequentially, but they cannot be turned on at the same time. Finding the switch in the dark is easy since it has an illuminated toggle. The toggle light turns on automatically when the switch is in its off position.

THE RUN

Defining a Run

A branch electrical power line consisting of a hot lead, a neutral, and a ground line is sometimes called a *run* and extends from the service box to the last electric box which can contain a receptacle or a switch. In some instances it may be just a junction box, not containing any components, but merely a storage place for the end of a branch, pending the installation of more electric boxes. This last box is an end of the line unit; that is, it receives the wiring of the run (Figure 6–13), but it does not have any wiring that continues elsewhere. Any electric box located after the service box and before the end of the run box is called a middle of the run. Middle of the run has no reference to physical distance, and there can be a number of middle of the run receptacles and switches. They are identified by the fact that they have both incoming and outgoing wiring.

Figure 6–14 shows a middle of the run arrangement consisting of an electric box and a receptacle. The box has two branch lines, one bringing power in and the other taking power out to some other electric box.

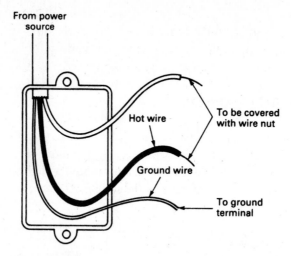

Figure 6–13. End of the run arrangement. Note that a power line comes in, but there is no power line going out.

End of the Run Possibilities

The end of the run of a power branch need not be terminated in a load. As indicated previously, the three leads, the black, the neutral, and the ground, can be individually capped with a wire nut and, for greater safety, can each be wrapped with electrician's tape. The three wires are then housed in an electric box, such as the rectangular type. The box is then closed with a solid cover plate.

There are two other alternatives. The end of the run can be terminated in a re-

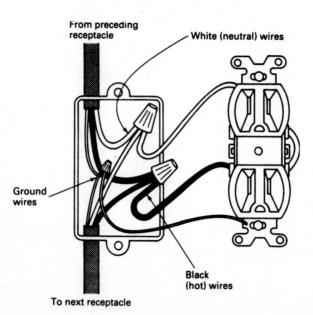

Figure 6–14. Middle of the run receptacle. The box is a nonmetallic type. Note that the box has a power input cable and a power output cable. The box itself has no ground terminal. The ground wire is connected to a grounding screw on the receptacle. The incoming and outgoing cables must be clamped to the box.

ceptacle. This receptacle can be regarded as hot, indicating that it is an unswitched type and is always available for a plug-in type of load. The other alternative is to connect the end of the run to a load, such as a lamp. If the end of the run is a lighting fixture, it can be turned on or off with its own built-in switch, possibly a pull-chain type.

There is a basic difference between a middle of the run installation and one that is at an end of the run. With a middle of the run, there is an entry feed from a branch line, and there is also a continuation of the line via an exit feed, leading to a following switch and then on to the next load. With an end of the run installation, there is no exit feed. An end of the run setup can always be modified to supply electrical energy to a continuation power line. For this reason it is advisable to have the end of the run unit in some convenient, accessible location.

Extending a Run[5]

In some instances a single length of wire may not be enough to complete the installation of a branch line. The obvious alternative is to measure the required distance and then to buy a new roll of wire. If the original length of wire has already been fastened into place, it may be easier to use an electric box for joining the extra wire. As a first step in doing this work remove the fuse for the branch line or open the circuit breaker.

After the wires have been securely fastened into place, the stripped ends of the wire can be joined with a pair of gas pliers. Join similar color-coded wires: black to black, white to white, and ground to ground. Replace the fuse or reconnect the circuit breaker. Use a neon-type test lamp or a volt-ohm-milliammeter set to read ac volts and check between the white and black leads. If the test is satisfactory, shut off power to this branch at the service box and cover the wire splices with wire nuts. For additional security, cover the wire nuts with plastic tape. As a final step, put a cover plate on the box using a pair of machine screws. Make sure the plate is fastened securely.

MISCELLANEOUS SWITCHES

Three-way Switching[6]

The purpose of a three-way switch is to control a single load, such as a light bulb, from two different locations. With such a switch and suitable wiring, it is possible to control a light located in a hall ceiling from downstairs or upstairs.

[5]*(NEC: Article 300–13; page 167)*
[6]*(NEC: Article 380–2; page 320)*

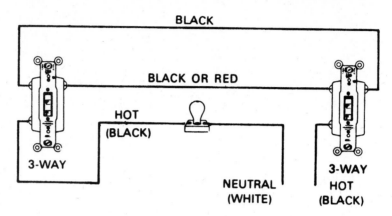

Figure 6–15.　Three-way switch controls a load from two locations. (*Courtesy Hubbell Incorporated, Wiring Device Division*)

Location of switch connections.　Note the connections to the three-way switches in Figure 6–15. This drawing shows that two of the connections are on the left side and one is on the right side.

The circuit for another pair of three-way switches is shown in Figure 6–16. Note that the switch connections are not the same as those in Figure 6–15. Before wiring a switch, be sure to consult the manufacturer's wiring instructions. These switches are equipped with a toggle that glows when the load is off. The point is not to assume that all switches are wired in the same way.

Alternative three-way switching system.　The preceding diagrams for three-way switching all used switches having three terminals. However, it is possible for such switches to have four or five terminals. Thus, the three-way switching diagram suitable for a three-terminal switch is not appropriate for one having more terminals. Figure 6–17 is a wiring diagram for five-terminal

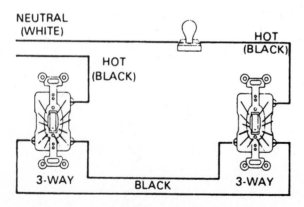

Figure 6–16.　Three-way switch equipped with glow-type toggle. (*Courtesy Hubbell Incorporated, Wiring Device Division*)

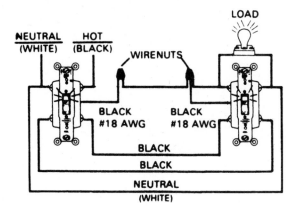

Figure 6–17. Alternative arrangement for a three-way switching system. (*Courtesy Hubbell Incorporated, Wiring Device Division*)

switches. The operation is the same. The idea behind all these circuits is to be able to turn a light on or off from two different locations.

Four-way Circuit for Light Control

The circuit shown in Figure 6–18 is used to control a single light and consists of a four-way switch and a pair of three-way switches. The four-way switching circuit can be used to control a light from three locations. This is a concept that can be extended. A load can be controlled from four locations by using two four-way switches between a pair of three-ways or from five locations by placing three four-ways between a pair of three-ways. Figure 6–19 shows some circuit arrangements for three-location control.

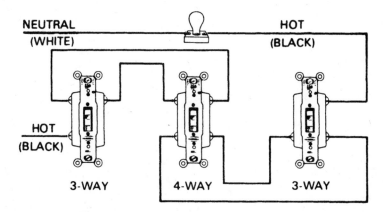

Figure 6–18. Four-way circuit for controlling a load from three locations. (*Courtesy Hubbell Incorporated, Wiring Device Division*)

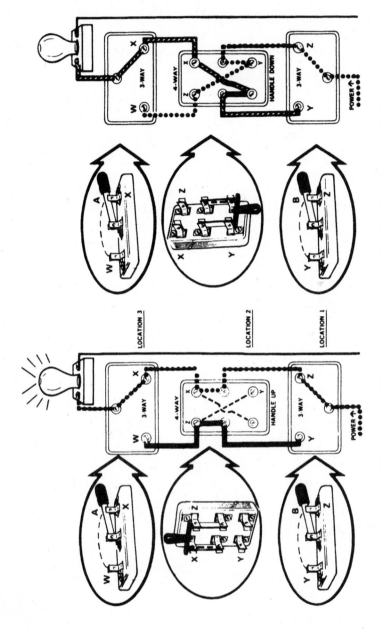

Figure 6–19. Two possible circuit arrangements for three-location control of a light. (*Courtesy Hubbell Incorporated, Wiring Device Division*)

How to Wire a Combined Switch and Receptacle

Although the unit shown in Figure 6–20 consists of a single-pole, single-throw switch and a polarized receptacle combined into one, they are independent of each other. As indicated in the drawing, the switch can be used to control a light, while the receptacle portion is always on and is independent of the switch. There are five terminals. A metal tab that ordinarily connects the two connections on the left side must be removed.

How to Install a Switch-controlled Receptacle

Instead of having a receptacle independent of its associated switch, as in Figure 6–20, it is also possible to have the receptacle switch controlled as in Figure 6–21. In this circuit the switch must be turned to its *on* position for the receptacle to be active. Note also that the electric box used in this drawing has three conduit connections, two across the top and one at the bottom. The usual number is two for middle of the run boxes. An end of the run box will have just one. The conduit at the top left leads to the branch power line, while the one at the right is the continuation of the same run. The conduit at the bottom is used to supply power to an additional receptacle.

How to Break into the Middle of a Run

It may sometimes be necessary to install a switch in the middle of a run. This can be done by tapping off an existing receptacle. Be sure to turn off the power to the

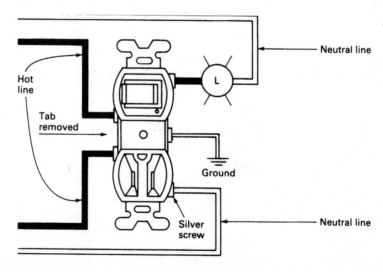

Figure 6–20. Combined switch and receptacle with each independent.

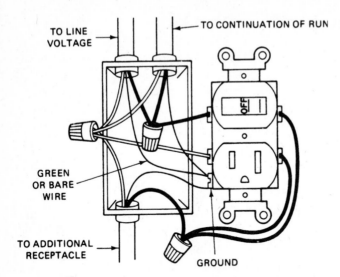

TO LINE VOLTAGE

TO CONTINUATION OF RUN

GREEN OR BARE WIRE

TO ADDITIONAL RECEPTACLE

GROUND

Figure 6–21.   Receptacle is on only when switch is turned on.

selected receptacle, remove its cover plate, and locate the incoming black, white, and ground wires. The easiest wire to use for the extension will be Romex, fastening it as indicated in Figure 6–22.

How to Control Two Loads with One Switch

Depending on the switch and the wiring method, two loads can be controlled either simultaneously or sequentially. The circuit shown in Figure 6–23 is used for turning on one load or the other, but not both at the same time. The switch is a single-pole, double-throw type. The two hot connections are on the left side, and the black leads of the two loads are connected to them. The neutral connections are on the right but are not wired to the switch, nor are they connected to it in any way.

This switch has three positions. When this switch is in its up position, the load represented by the upper light turns on. When the switch is in its bottom position, a lower load turns on.

Figure 6–22.   Connectors are available for clamping plastic-sheathed cable to electric box. Note connection of ground wire to grounding clamp in the box. Plastic-sheathed cable is easier to work with than conduit or BX.

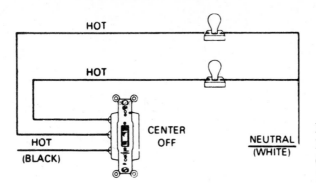

Figure 6–23. Wiring for controlling two loads with a single switch. (*Courtesy Hubbell Incorporated, Wiring Device Division*)

When the toggle is in its center position, both loads are turned off. The toggle is equipped with a light to indicate when it is in its *on* position.

HOW TO INSTALL ADDITIONAL WIRING[7]

With the growth of appliance use in the home, it may become necessary to install new wiring. An easy way is to use cube taps and receptacle strips, but a more desirable technique would be to do some rewiring. Rewiring can consist of upgrading existing power branch lines, possibly replacing BX with Romex cable, using grounded receptacles instead of the nonpolarized, nongrounded types, by using a heavier-gauge wire to carry more electrical loads, and replacing a fuse box with a circuit-breaker type. If a heavier-gauge wire is used, it must continue throughout the entire length of the run plus any branches from that run.

Updating a wiring system can be elaborate or it can be done on a one step at a time basis. Additional wiring can begin at the last receptacle in a run, but new wiring can also be tied in at the electric box of a ceiling light. The last receptacle and the ceiling box are good choices since both are two wire operated. Before starting any new wiring, plan to make the new run as short and as convenient to install as possible.

In all instances a new run, a continuation of an existing run, starts in an electric box. This electric box may be an existing one or a new one placed immediately adjacent.

A new run can be started from an end-of-the run receptacle using the two available screws located one on either side of the lower part of the receptacle. Before doing so, make sure that power to the receptacle is turned off.

[7](*NEC: Article 210–4; page 53*)

Adding a New Receptacle from a Junction Box

It is often desirable to make connections to the wires in a junction box, although this may be done from any receptacle. The problem with doing so, however, is that it adds wires to an electric box that may already have the maximum allowable number of wires. The disadvantage of using a junction box is that it may not be conveniently located for putting in a new receptacle.

The wires for this installation, consisting of a neutral, a hot lead, and a ground lead, are brought in from a junction box. After the connections have been made in the junction box, be sure to replace its cover plate.

SMOKE ALARMS[8]

An electrical fire can be accompanied by smoke due to burning insulation, and it can have a strong acrid odor. As an installation priority, install a smoke alarm in the area around the bedrooms. Try to protect the exit path, as the bedrooms are usually farthest from an exit. Figure 6–24 shows smoke-detector locations for single- and multiple-floor homes.

Position additional detectors to protect any stairway as they act like chim-

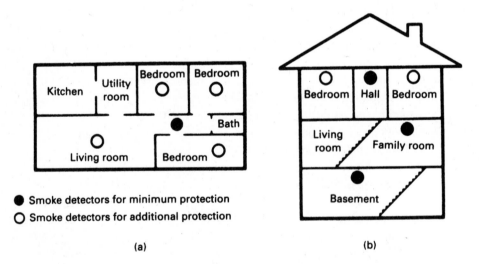

(a)   (b)

Figure 6–24.   Positioning of smoke alarms in (a) a single-floor home and (b) a multistory home. (*Courtesy National Fire Protection Association*)

[8](*NEC: Article 760; page 834*)

neys for smoke and heat. Place at least one detector on every floor level and also in any area where a smoker sleeps or where electrical appliances are operated in sleeping areas.

Smoke, heat, and other combustion products rise to the ceiling and spread horizontally. Mounting the detector on the ceiling in the center of the room places it closest to all points in the room. Ceiling mounting is preferred in residential construction. However, in mobile homes, wall mounting on an inside partition is required to avoid the thermal barrier that may form at the ceiling. When mounting a detector on the ceiling, locate it at a minimum of 6 inches from a side wall and 2 feet from any corner. When mounting a detector on a wall, use an inside wall with the detector a maximum of 12 inches below the ceiling and at least 2 feet from any corner.

There are two basic types of alarms, battery operated and those that receive operating power from a branch line. Both have advantages and disadvantages.

The battery type should be checked monthly for performance since the battery will not last indefinitely. If the battery stops working, check it with a battery tester since the fault may not be in the battery. The disadvantage of the power line-operated smoke alarm is that it requires connection to a receptacle.

Do not locate a smoke detector in these places:

1. The kitchen. Smoke from cooking might cause an unwanted alarm.
2. The garage. Products of combustion are present when an automobile is started.
3. In front of forced-air ducts used for heating and air conditioning.
4. In the peak of an A-frame type of ceiling.
5. In areas where temperatures may fall below 40°F or rise above 100°F.

False Alarms

A well-constructed detector will be designed to minimize false alarms. Smoking will not normally set off an alarm unless the smoke is blown directly into the detector. Combustion particles from cooking may set off the alarm if the detector is located close to the kitchen. Large quantities of combustion particles can be generated by different cooking processes. If the smoke alarm does go on, look for fires first. If no fire is found, check to determine the reason for the alarm.

NFPA Recommendations

The National Fire Protection Association Standard No. 74 on Household Fire Warning Equipment, 1975 edition, recommends that one of the minimum levels of protection given in Table 6–1 be provided in each residential occupancy.

TABLE 6–1. LEVELS OF PROTECTION

Level	Detection Equipment Required	Where to be Installed:	
		Smoke Detectors	Smoke or Heat Detectors
1	One or more smoke detectors plus additional smoke or heat detectors	To protect each separate sleeping area and at the head of each basement stairs	All other major areas and rooms of the living unit including any basement.
2	One or more smoke detectors plus additional smoke or heat detectors	To protect each separate sleeping area	Basement, living room, bedrooms, attic, furnace (utility) rooms
3	One or more smoke detectors plus additional smoke or heat detectors	To protect each separate sleeping area	Basement, kitchen, living room, furnace (utility) rooms
4	One or more smoke detectors	To protect each separate sleeping area and at the head of each stairway to occupied areas	Not applicable

SWITCHED RECEPTACLES

There are a number of wiring locations for switches. They can be toggle operated in-the-wall units, on-the-wall types, small light-weight thumb-wheel types that are part of a lamp cord or built into an electronic component such as a radio or TV receiver. In all of these the switch is a single, independent component.

There is another category that consists of a switched receptacle. In this example one of the receptacles in a duplex type is replaced by a toggle operated, single-pole, single-throw switch. Depending on the wiring the remaining receptacle and the switch can be dependent or independent of each other. When independent, the receptacle is always live and a plugged-in component will receive power the moment it is plugged in. The switch can operate any selected device.

Another arrangement that can be made consists of the switch controlling its associated receptacle. In this case the switch controlled receptacle must have the toggle turned to its on position for the receptacle to receive power.

The three parts of Figure 6–25 show applications of this component. Part (a) shows the switched receptacle being used to control a light only while the receptacle can be used by grounded appliances. This arrangement is helpful if an appliance is to be operated in a dark area. The light comes on automatically when the switch is set to its on position.

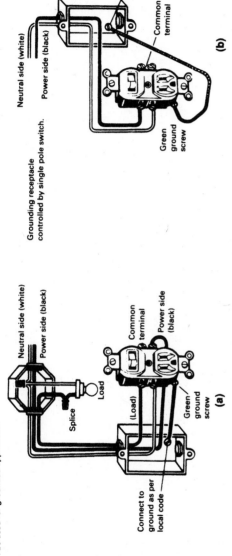

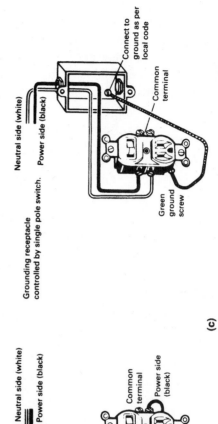

Figure 6–25. Methods of wiring a combination switch and receptacle. (*Courtesy Leviton Manufacturing Co., Inc.*)

In part (b) the switch is used to turn power on or off to any device plugged into the receptacle.

Part (c) shows the switch and receptacle on separate circuits. The switch-receptacle assembly has a fin on the right side. It will be necessary to remove this connecting fin, a small bit of metal, to enable the switch and the receptacle to operate independently. To do this, insert the blade of a screwdriver into a slot on the fin and bend it back and forth until it breaks off.

Chapter 7

Practical Wiring

PART II[1]

Possibly the simplest wiring consists of a single switch controlling just one light. But no matter how simple or complex, all wiring consists of tapping into a branch line as the source of power. Even though a number of branches are used, all loads are in parallel with each other. Switches, circuit breakers, and fuses are always in series with those loads, and while the total load current flows through them, they do not constitute a load. Consequently, inserting these components into the line does not have the effect of increasing the current, but is simply for the purpose of current control.

Not all wiring drawings supply the same amount of information, and some are more detailed than others. After the connections are made, a component such as a switch is fitted into a rectangular box. If the box is metal, be careful to make sure that none of its side-mounted machine connecting screws touch the sides of the box. The clearance between these screws and the box is small.

[1](*NEC®: Article 300, page 159*)

BRANCH CONNECTIONS

Connections to a branch power line are either end of the run or middle of the run. End of the run can be recognized by the single power-line entry port and an absence of an output power cable. A middle of the run has at least a single entry and an exit cable, but can have more.

End of the Run Junction Box

An end of the run junction box is the terminal point of a power cable. The cable that enters this box consists of the black lead, white lead, and ground lead. The black and white leads have their stripped ends covered with wire nuts, while the ground lead is joined to the ground terminal of the box, assuming the box is metal. If the box is plastic, the stripped end of the ground wire can also be covered with a wire nut.

End of the Run Receptacle

The end of the run electric box can contain a receptacle wired as indicated in Figure 7–1. This receptacle can have side, front, or rear connecting terminals. The wires in this drawing are side connected. Since the receptacle is not switch controlled, it is always supplied with voltage. The box used in Figure 7–1 is metal since the ground wire is connected to a grounding screw on the back of the box. The ground wire is

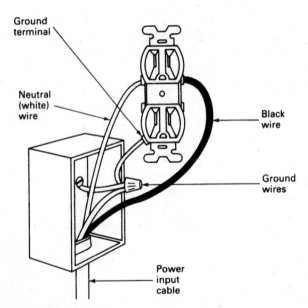

Ground
terminal

Neutral
(white)
wire

Black
wire

Ground
wires

Power
input
cable

Figure 7–1.   End of the run receptacle.

also joined to a grounding terminal on the receptacle. This grounding terminal is internally wired to the round holes shown at the sides of each of the two receptacles. This supplies automatic grounding of any three-prong plug used by an appliance.

A feature of this receptacle is the fact that the connecting screws on the left side are joined by a strip of metal. The screws on the right side are similarly joined. The advantage of this technique is that it simplifies the wiring. If these metal strips are omitted, then only the upper part of this duplex receptacle will be powered.

The end of the run receptacle can supply immediate power to a plug-equipped load. However, it is also possible for a load to work directly with an end of the run setup, as indicated in Figure 7–2. The load in this case is a lamp. This load can be equipped with a built-in pull-chain switch (not shown in the drawing) or else can be controlled by a switch (Figure 7–3) preceding the end of the run installation. However, the switch may not be necessary if the lamp fixture has its own built-in pull-chain switch. The lamp fixture may also have one or two built-in receptacles.

End of the Run Control

The end of the run setups that have been described can be controlled by the use of a SPST switch set up in the middle of the run as indicated in Figure 7–3 and then connected to the end of the run. By doing this, it is possible to have a controlled

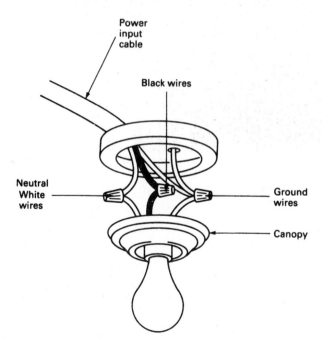

Figure 7–2. End of the run light fixture.

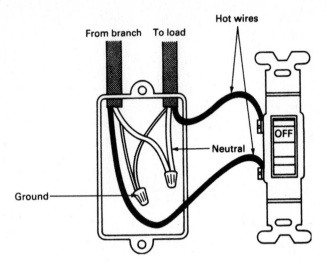

Figure 7–3.   Connecting a switch for end of a run.

receptacle or a controlled lamp. In some instances it may be desirable to have the capability of turning the receptacle off, as when a lamp is used that does not have a built-in switch.

How to Connect a Middle of the Run Receptacle

The electric box used in Figure 7–1 is a metal type. This is evidenced by the fact that the ground wire is fastened to the box. The middle of the run box (Figure 7–4) is also metal for the same reason. The electric box in this drawing can be recognized as middle of the run since it has an input and an output port. The input port contains a cable from a preceding power source, while the output port has a power cable that continues on to a following receptacle.

Middle of the Run Load

The middle of the run can also have a directly connected load, as in Figure 7–5a. In this case the load is a lamp, and the easiest arrangement is to use a porcelain type that has a built-in pull-chain switch. Otherwise, it will be necessary to control this load with a separately installed switch.

How to Setup Twin Receptacles

Figure 7–5b shows a pair of identical receptacles mounted in a metal box. Connect the white wire to the white terminal and the black wire to the brass-colored terminal. This ensures that the prongs of the mating plug will be connected properly. The connecting cable is a three-wire type containing a white wire, a black

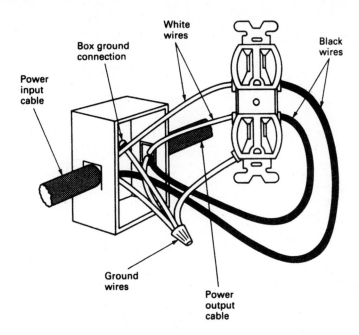

Figure 7–4. Middle of the run receptacle.

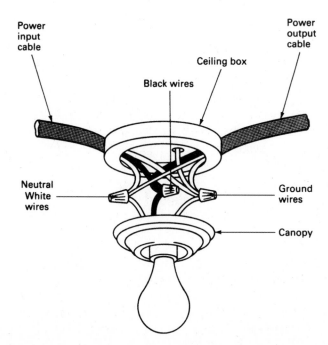

Figure 7–5a. Middle of the run fixture.

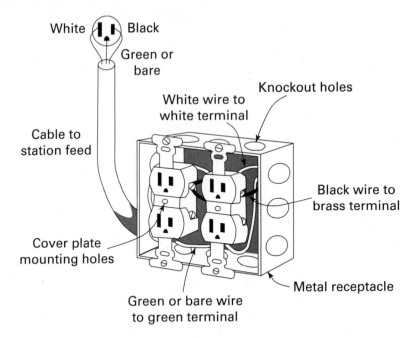

Figure 7–5b.   Pair of identical receptacles mounted in a box.

wire, and one that is green. In some cables the green wire is replaced with one that is bare. The box should have a number of through holes to accommodate machine screws for fastening the box to a stud.

Wiring a Dimmer Switch[2]

All toggle switches, regardless of type, are go, no-go components. They have only two possible positions, on or off. A dimmer switch, however, not only has these two positions, but a large number of variable positions between these extremes.

A dimmer switch can be considered a modified toggle type having two operating possibilities. One is its use as an on-off switch. The other is a variable control that governs the amount of current flow through the switch portion. Thus, it is possible to have any level of light from a controlled bulb, depending on the setting of the variable control.

Dimmer types.[3]   Dimmers are made to operate with either incandescent lights or fluorescent fixtures. One type of dimmer has a high-low control that supplies two levels of illumination, full brightness or half-brightness. When the dimmer is in

[2]*(NEC: Article 520–25; page 602)*
[3]*(NEC: Article 520–25 b–d; page 603)*

its low position, a silicon rectifier is put in series with the white lead connected to the lamp. The rectifier reduces the current flow by about 50%, and so the light becomes dim. In its high position the switch in effect removes the rectifier from the circuit, allowing the lamp to receive a full current, and so the lamp glows at full brightness.

There are a number of advantages in using a dimmer. The nearest equivalent is a three-way light, but this supplies only three levels of illumination. The dimmer has a much greater range. The dimmer is more economical to operate. If the controlled bulb is rated at 100 watts, the dimmer can effectively cut that rating to 30 watts or 40, or any rating below 100.

A dimmer can be used to control just a single bulb or a number of them. Consequently, dimmers have maximum wattage ratings, and this should not be exceeded. If the fixture to be controlled has 8 bulbs with each having a 40-watt specification, then the minimum rating of the dimmer should be $8 \times 40 = 320$ watts.

Installing a dimmer.   Dimmers are installed in the same way as ordinary toggle switches. The dimmer control is usually housed in a small sealed box. Remove the existing switch, connect the wires to the two terminal screws on the dimmer, and then insert the dimmer into the electric box.

One type of dimmer is equipped with a shaft that is knob controlled. This shaft protrudes through the wall plate and has a double function. When it is first rotated, it turns the light on, and as the rotation is continued, the light will vary from dim to maximum brilliance. Dimmers are made for use with incandescent lights or with fluorescents, but they are not interchangeable.

Since the dimmer is a switch, it can have the characteristics of and the names used for switches. Thus, a single-pole dimmer is used for controlling a single light fixture. To use a dimmer having two-location control, use a three-way dimmer in place of any one of the three-way switches in a two-way setup. A socket dimmer is still another type. Remove the light bulb and replace it with the dimmer and then screw the bulb into the dimmer.

A dimmer can cause electrical interference, which may produce radio receiver noise and cause picture problems with TV sets. The problem can be minimized or eliminated by moving the radio or TV set as far from the dimmer as possible or by using a different branch line for the entertainment components.

Fan Speed Control[4]

The purpose of a fan control is to permit a smooth adjustment of its speed up to its maximum rotation. Some fans are equipped with fan controls, but usually this is a two- or three-speed device.

[4]*(NEC: Article 680–41b; page 769)*

The fan control in Figure 7–6 is a solid-state component, and its connections are the same as those for a dimmer. The control is determined by the operating power requirements of the fan, as, for example, 120 volts, 5 amperes. This is the maximum specification for a fan for a selected control.

The fan control replaces the usual SPST switch. After making the switch selection, remove its fuse from the service box or open its circuit breaker. Remove the switch wall plate and then take out the switch. It will have two black wires and a ground lead. Disconnect these. If the old switch has a ground wire normally connected to a green-colored hex head screw, fasten it to the metal box. If the box is plastic, fasten the ground wire to either one of the fan control mounting screws. Connect the two black wires to the corresponding black wires of the fan control. Use wire nuts when making the connections. Mount the control in the electric box and restore power.

Turn the knob clockwise to turn the fan on and continue in this direction to reduce the fan speed. Note that the maximum fan speed is obtained when the knob is in its maximum counterclockwise position.

How to Connect an Electric Range or Dryer[5]

The voltage used by an electric range or dryer is commonly 240 volts. Unlike other appliances in the home, a range, dryer, or washing machine will not share its branch line, and so there will be separate power lines going to these components. This will consist of a four-wire cable of No. 6 wire.

Inside the housing for the appliance, near the rear, will be a terminal for connecting three wires (Figure 7–7a). The connection on the left is for a black wire (B) and that near the right is for a wire color coded red (R). Both are hot leads and measure 120 volts with respect to the neutral (G) wire and 240 volts from the red wire to the black. There is also a fourth wire not shown in the drawing. This is the ground wire, and it will be fastened to the metal housing of the appliance and to the ground terminal of the plug.

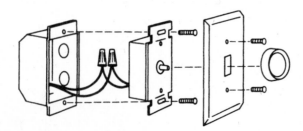

Figure 7–6.   Fan speed control. (*Courtesy General Electric Co., Wiring Devices*)

[5](*NEC: Article 422–17; page 383*)

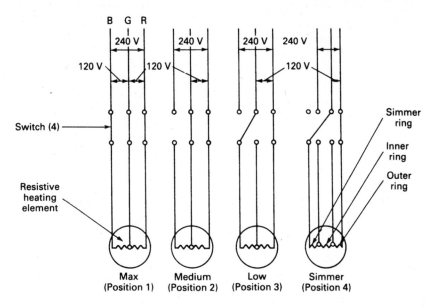

Figure 7–7a. Wiring for an electric range. The ground lead is not shown in this drawing but it is part of the power cable.

A dryer may use either 120 or 240 volts, depending on the function of the unit at a particular point in its cycle. The dryer can have a power demand of 4200 watts, while some special units having a high-speed capability may demand as much as 8500 watts. Prior to installing a high-power appliance, check with your local power utility. A special power line may need to be installed. In some areas the wiring for current-demanding machines may need to be done or supervised by a licensed electrician.

Manufacturers of electrical equipment often supply wiring diagrams to accompany their products. These drawings can vary in detail and some are easier or more difficult to understand. Compare the drawing in Figure 7–7b with that in 7–7a. That in Figure 7–7c shows essentially the same details but also includes the wiring of the receptacle.

For a current-hungry component such as a dryer, it is recommended that a dedicated line be used, that is, a separate circuit from the entrance panel. On a 240-volt ac circuit the two line wires should be protected with a 30-ampere circuit breaker. Do *not* use a circuit breaker on the neutral line. Recommended wire sizes for 240-volt usage, up to 40 feet, is No. 10 wire, and No. 8 wire for up to 60 feet. Note that the data here are given for a particular dryer and that the circuit breakers are 30-ampere types (Figure 7–7c).

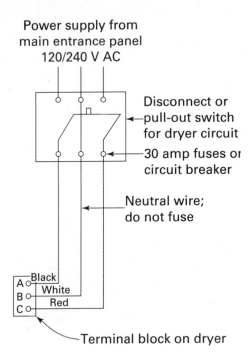

Power supply from
main entrance panel
120/240 V AC

Disconnect or
pull-out switch
for dryer circuit

30 amp fuses or
circuit breaker

Neutral wire;
do not fuse

A
B
C

Black
White
Red

Terminal block on dryer

Figure 7–7b.   Type of circuit
wiring for a dryer.

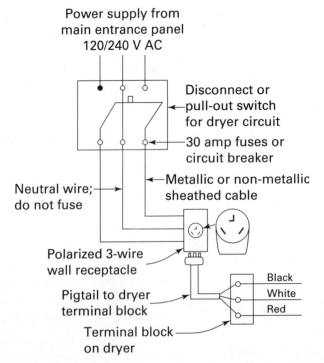

Power supply from
main entrance panel
120/240 V AC

Disconnect or
pull-out switch
for dryer circuit

30 amp fuses or
circuit breaker

Metallic or non-metallic
sheathed cable

Neutral wire;
do not fuse

Polarized 3-wire
wall receptacle

Pigtail to dryer
terminal block

Black
White
Red

Terminal block
on dryer

Figure 7–7c.   More detailed diagram for the same dryer.

CIRCUIT SURVEY[6]

The circuit survey is an easy and inexpensive way of determining the operating safety of each power branch in a building, no matter what kind it may be. It will also be found to be helpful in the event of an electrical fault.

The service box may have a ruled sheet on its inside front panel. This sheet will have a numbered listing, and its entries will identify each power branch. If there is not enough room on the sheet for all the data that should be listed, use a different sheet pasted in place of the original.

On this sheet list each branch line, indicating the rooms or areas that it services. Give each circuit a number and indicate the types of appliances used in each and their wattage rating. Do this even if the appliances are used only intermittently. It would also be well to indicate the gauge number of the wire in each branch and its ampacity. As an example, the branch might use No. 14 gauge wire, have a 1650-watt limit, and be protected by a 15-ampere breaker. Its present power load is 1125 watts. This is very close to the power limit of this branch, and it would not take many additional appliances to create an unacceptable electrical situation.

There are several solutions to the problem of a branch circuit that is close to its overload point. One is to rewire the branch with a heavier wire; in the preceding example we would replace the No. 14 wire with No. 12 gauge. If the service panel has an unused fuse or circuit breaker, install another branch or replace the service panel with one having a larger number of available fuses or breakers. Replacing the original fuse or breaker with one having a higher current rating is not a solution.

The wiring for the new branch (Figure 7–8) can consist of a cable containing two hot wires, one black and one red, a neutral, and a ground wire. Connect the white wire to the neutral strip of the service panel. Attach the two hot wires to the corresponding power take-off lugs. The cable leading from the service panel will now supply 120 volts between the neutral and either black or red wire, or 240 volts between the black and red wires. The cable leading away from the service box need not include both wires. Generally, both will be used if the cable is to service a heavy-current device such as an electric stove, clothes washer, or dryer. It may be required as a dedicated line for the first-time installation of these appliances, in which case both hot lines will be used.

Data are not always entered on the service panel for new constructions. If not, this should be done by the occupant. This is easy enough to do by opening the breakers one at a time, or removing the fuses, and learning which receptacles or ceiling loads are affected. Information concerning the wire gauge used for each

[6]*(NEC: Article 210; page 52)*

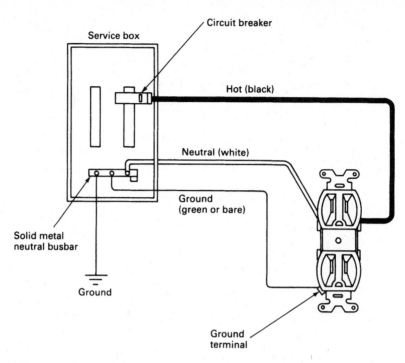

Figure 7–8.   Connection of new branch line to a circuit breaker in the service box.

branch can be obtained from the wiring contractor or, if Romex is used, from data printed on the wiring insulation.

HOW TO INSTALL A DOORBELL, BUZZER, OR CHIMES

Although doorbells and buzzers were once battery-operated devices, they now work from the ac power line via a transformer having a 120-volt primary and a 10-volt secondary. While 10 volts is commonly used, some doorbells are intended to operate from 16 volts and some from 26 volts. Some transformers are capable of supplying all three voltages using a split secondary. A transformer supplying 10 volts only can easily be recognized since it will have only two secondary terminals. The multiple-voltage output transformers have three secondary terminals. The multiple-tap transformer can be used to supply 16 volts for chimes or 10 volts for a doorbell.

The primary of the transformer (Figure 7–9) remains permanently connected to the line voltage, supplying what is known as a magnetizing current. The doorbell is connected to the secondary via a front-door push button using the cir-

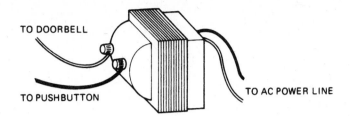

TO DOORBELL

TO PUSHBUTTON

TO AC POWER LINE

Figure 7–9. Transformer for operating a door bell. The wires connected to the power line are from the primary winding. The two wires at the left are connected to the secondary.

cuit appearing in Figure 7–10. The push button is actually a type of single-pole, single-throw switch. Note that the circuit does not contain a fuse. The arrangement depends for fusing on the branch line to which it is connected. The transformer can be mounted directly on the cover plate of an electric box with the wiring inside the box to power line leads.

Wiring for the primary and the secondary can be No. 16 or No. 18 gauge two-wire Romex. Bell wire is sometimes used, but is not recommended since its insulation is easily pushed back.

Wiring Front and Back Doorbells or Buzzers

An additional doorbell or buzzer can be installed at the rear door of a home. A buzzer is used to supply a different audible signal to enable quick identification.

Figure 7–11 shows the circuit. It uses a single transformer to supply voltage

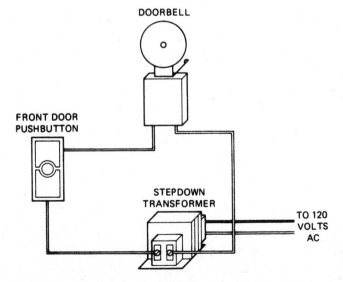

DOORBELL

FRONT DOOR
PUSHBUTTON

STEPDOWN
TRANSFORMER

TO 120
VOLTS
AC

Figure 7–10. Circuit for operating a door bell.

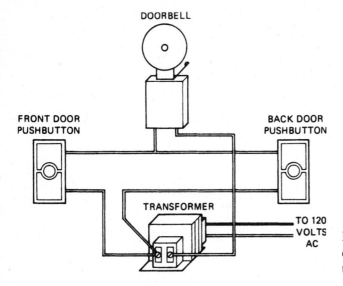

Figure 7–11. Single doorbell operated by front and rear push buttons.

for a single doorbell. When depressed, either push button will supply current to the doorbell. The push buttons are wired in parallel.

The circuit in Figure 7–12 shows the arrangement when a buzzer is to be used for the rear door instead of a doorbell. In this circuit and in the preceding one, both doorbells, or doorbell and buzzer, require the same operating voltage.

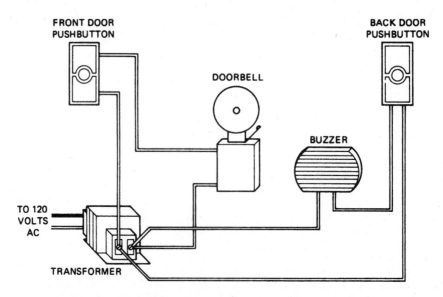

Figure 7–12. Doorbell identifies front door; a buzzer identifies the back door.

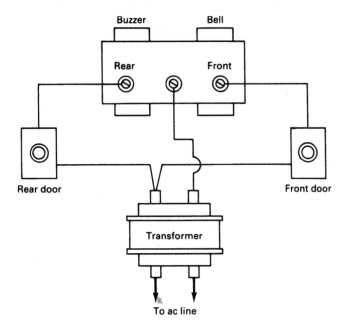

Figure 7–13.   Combination bell and buzzer circuit.

Wiring a Combination Bell and Buzzer

The wiring of a front-door bell and a rear-door buzzer can be simplified by using a combination bell and buzzer unit. The circuit arrangement is shown in Figure 7–13. The combination type can be easily recognized since it has three connection terminals.

Wiring a Door Chime

The difficulty with a door bell or buzzer is that these are unattractive units and so can present an installation problem. Also, the sound they make can be unpleasant. An alternative is the use of chimes.

Unlike doorbells and buzzers, which do not supply a melodic tone, chimes can produce two, four, or as many as eight tones. Some play a tune. Also, separate notes can be played for either the front or rear door for easy identification. The circuit for a two-tone chime for front and back doors is shown in Figure 7–14. The chime in this drawing has three terminals. The transformer must be able to supply 16 volts and may be somewhat larger and heavier than that used for doorbells and buzzers.

Some chimes use the same tone for both front and back doors, with two identical notes for the front but just one for the rear.

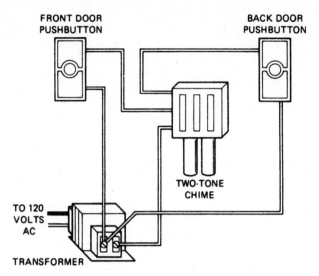

Figure 7–14. Circuit for two-tone chime.

Some homes are equipped with three entry doors: front, back, and side. For these, the circuit shown in Figure 7–15 can be used. Three push buttons are required, one for each door. The front door chime has two tones; each of the others has a single tone. The single tones can be different to permit quick identification of the door requiring attention.

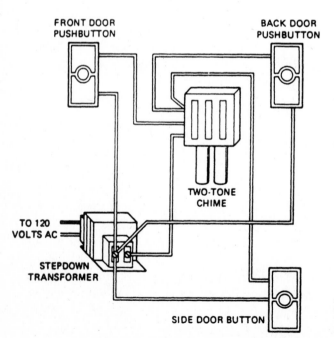

Figure 7–15. Circuit arrangement for using three push buttons.

Door Opener

A door opener is a device for releasing a catch that permits the door to be opened from the outside. It works the same as a doorbell, buzzer, or chime. It is push button operated with that control located in some convenient place, such as a kitchen. Pushing the button allows a small current to pass through a catcher type of latch positioned in a door lock. The current releases the latch, allowing the door to be opened externally. Once the door closes, the latch fastens automatically. The latch is a vibrating type, and the noise itself is an alerting signal.

GROUND-FAULT CIRCUIT INTERRUPTER (GFCI)[7]

Unlike ordinary circuit breakers, which respond to what is usually a heavy overload current, the GFCI (Figure 7–16) turns off power in response to a small leakage of current, generally from 3 to 5 milliamperes (0.003 to 0.005 ampere). It might seem that such a small current could not possibly be unsafe, but it can result in death.

Both the black and white leads in a power circuit are current carriers and both must carry the same amount of current. If two hot leads are used, a red and a black, the current flowing through the neutral wire must be the sum of their currents. No current is ever lost, and the current entering a service box must be equal to the current leaving it, and vice versa.

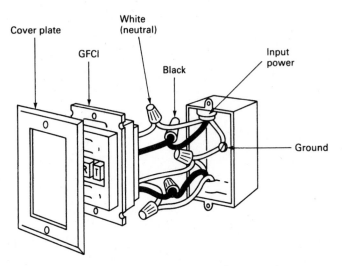

Figure 7–16.   Wiring of a GFCI.

[7](NEC: Article 210–8; page 57, 215–9; page 68)

A GFCI makes a comparison of these currents, and if there is a difference of just a few milliamperes, it works as a circuit breaker and does so in about $\frac{1}{40}$ second. This is fast enough to prevent bodily damage.

If the GFCI is inoperative, which is evidenced by none of the appliances working in the protected branch, it can be tested by pressing its reset button. If electrical service is restored in about 1 second, then the leakage current may have been a momentary intermittent, and the branch circuit can be used. If the restoration of power lasts about 1 second with the GFCI tripping again, there is a current leakage fault that requires attention and repair.

The electrical fault is somewhat localized if there is a separate GFCI at the beginning of each branch. If an overall GFCI is installed in the service box, then there is a problem of determining which electrical branch is affected.

Location of the GFCI[8]

There are various possible locations for a GFCI; one is at the service box, while others can be positioned in one or more receptacles.

In the service box a GFCI breaker can be used to replace the usual 120-volt breaker, or several can be used to control a number of branches. The GFCIs are rated for 15 to 30 amperes, and so these can be used for two different current branches. GFCIs are also designed for 120- and 240-volt circuits.

In any branch, if the difference in current flow in the hot and neutral leads is more than about 3 milliamperes, the GFCI breaker will open. The GFCIs are equipped with a test button. Push it, and if the GFCI is in working order it will trip. Reset it by pushing its toggle back into place.

Another type of GFCI can be had that replaces the usual polarized duplex receptacle. If it is installed in the first electric box in a run, it will protect all the following receptacles in that run. For new installations, a NEC requirement is that all receptacles within 6 feet of the kitchen sink must be GFCI protected. Also, GFCIs are required in basements, garages, and bathrooms. All receptacles located outdoors must have GFCIs. GFCIs are equipped with test buttons. When these are pushed, a reset button will come out of the unit. Push the reset button to reactivate the GFCI.

Types of GFCIs

A number of different GFCIs are available for different applications. The best way for new construction is to have the unit installed in the service box. It combines ground-fault protection with overcurrent protection against overloads and

[8](NEC: Article 210–8; page 57)

short circuits. Two different units can be used: single pole for 120-volt protection and two pole for 120/240-volt systems.

For existing homes where the service box is a fuse type or cannot otherwise accommodate a GFCI, install the unit in an electric box to replace an existing receptacle. In this group there are two types; the first protects only components that are directly plugged in, and the second is a feed-through type that supplies protection to its own area of use and all other receptacles downstream on the same branch line. In either case the unit works as a three-prong, duplex polarized receptacle.

Probably the easiest to install is the plug-in GFCI. Simply insert the GFCI at the point of use. It also comes equipped with a duplex receptacle.

The surface-mount GFCI is equipped with its own electric box and can be attached to a wall, stud, or any other location where on-wall wiring is used. For example, it would be suitable in a basement, near a workbench, or in a garage or barn.

For outdoor work, use an extension cord with a built-in GFCI.

Installing a GFCI

Remove the existing receptacle but first, turn off the power by removing the fuse or turning the circuit breaker to its off position. As a further check, use a voltage tester such as a neon lamp type to make sure the receptacle has been inactivated. Then remove the receptacle. The GFCI will have three wires, black, white, and ground (either bare or with green insulation). These wires may be marked "line," an indication that they are to be connected to the power lines entering the receptacle. Connect the similar colors of the power cable and of the GFCI. Use wire nuts to cover the connections. Install the GFCI in the electric box and use its pair of machine screws to fasten it in place.

Not all GFCIs will fit into a standard-sized electric box. It will help to measure the dimensions of both before moving ahead with the work. The GFCI has its own cover plate; that used for the replaced receptacle will not be suitable.

LINE VOLTAGE SURGES AND SPIKES[9]

Line voltage is not constant and can vary from about 110 volts to as much as 125. The amplitude of this voltage depends on the generators used by a utility, their regulation (their ability to maintain constancy of voltage under load), the amount of load, and the distance of the load from the generator. The frequency of the voltage supplied by these generators is remarkably constant, and while it may vary by a half-cycle during any 12-hour period, it averages 60 hertz in a day.

[9]*(NEC: Article 280; page 156)*

The voltage supplied by a utility, when that voltage gets beyond the service box, is subject to two forces over which the electric utility has no control. One of these is an induced voltage and the other is a feedback voltage. The induced voltage is produced in branch lines by stray magnetic fields. The feedback voltage is the voltage fed back into branch lines by all types of electrical and electronic devices.

As a consequence, the current flowing in a branch line is afflicted by voltage surges and voltage spikes. There is a difference between these two consisting of time duration and amplitude. A voltage spike can exist for part of a second, but its strength can be as high as 6 kilovolts. The voltage surge lasts longer, but its amplitude is less, somewhat on the order of 1 kilovolt. Both are capable of damaging equipment via the power line.

Electrical noise from the power line can cripple sensitive electronic equipment. It is known as electromagnetic interference (EMI) and is caused by feedback into power lines by motors and fluorescent fixtures. While it is desirable to have only pure sine-wave ac, that sine wave can be accompanied by RFI (radio frequency interference), EMI and electrical noise. One or all of these electrical impurities can damage VCRs, TV sets, computers, and other sensitive electronic devices. They can wipe out VCR settings and cause a microwave clock to blink on and off and a computer to malfunction. The cure is to use a surge suppressor or an uninterrupted power supply system (UPS).

The voltages that afflict the power-line voltage come under the general heading of transients. They can occur when a motor switches on or off, including those in a refrigerator, washing machine, oil burner, and power tools, or lightning can produce voltage surges that can travel along a power line. The result can be immediate damage or gradual deterioration of electronic devices, ultimately resulting in equipment failure that is hard to diagnose. In the case of a surge due to lightning, the result could be a fire in a line voltage-operated device.

There are three types of transient voltage suppressors: the silicon avalanche diode (SAD), the metal oxide varistor (MOV), and the gas tube. The first two of these are voltage-clamping devices; as such they impose a limitation on the ability of the branch line to let them by. The gas tube has a negative resistance, a technical way of saying that it acts like a short circuit for the voltage transient.

Surge suppressors are easy to install for they can be plugged directly into a receptacle. They also are built into multireceptacle power strips. Thus, a strip containing six receptacles will have all of them surge protected. A surge suppressor offering protection of up to 6 kilovolts should be adequate. The unit may be equipped with an excessive surge alarm to warn if the suppressor unit has an electrical disturbance in excess of its protection capability.

UPS (uninterrupted power supply) systems are made primarily for commercial use to protect computer systems and mainframes. They are also available in desktop models for PCs (personal computers).

Figure 7–17 is the circuit diagram of a surge suppressor using three metal oxide varistors, identified as TS in the diagram. To install the unit, select a receptacle for replacement by the surge suppressor. Make sure power is turned off and then remove the receptacle. Connect the black lead from the input power branch to the terminal marked B, the ground wire to G, and the neutral to N. This unit has a push-to-reset button that includes a neon glow lamp. Normally, this lamp will be lighted to indicate that the surge suppressor is working. If the lamp does not work, it is an indication that one or all of the varistors are worn out from excessive voltage and that the suppressor should be repaired.

This suppressor is equipped with a duplex polarized receptacle. The circuit includes a filter to supply protection against electrical noise riding in on the power line. Some suppressor modules are also available with a single receptacle.

Do not use a surge suppressor with a nonpolarized receptacle. If these are the only kind available, use an adapter and connect the lead of the adapter to the center screw of the face plate. This screw must be grounded. Do not use the surge suppressor outdoors or in wet, oily, or high-temperature conditions. Do not use the suppressor across a 240-volt line.

Surge Suppression for Computers

A basic computer system consists of three units: a central processing unit (CPU), a monitor, and a printer. These are all power-line operated and should work from the same power source. If the system also includes power-operated add-ons, they should be connected to the same power receptacle. To avoid unsightly cube taps use a multiple power strip that is surge-suppression equipped.

The surge suppressor should carry a UL label but this is not enough. The label should also display a 1449 listing. The amount of voltage to be suppressed will vary depending on how it was produced. If generated by lightning and if it

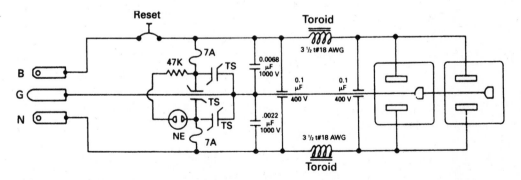

Figure 7–17. Circuit diagram of a surge suppressor. (*Courtesy Radio Shack, a Division of Tandy Corporation.*)

finds easy access to the computer, survival by the computer is unlikely. A much more possible surge voltage source is one generated by some indoor motor-operated appliance, such as a garbage disposal unit, a dishwasher, a washing machine, or its accompanying dryer.

The line voltage input to the computer may be 120 volts but the computers circuitry drops this to about 5. It doesn't take much of a percentage increase produced by a voltage spike to result in a damaging increase.

The surge protector should come with a label listing the unit's electrical characteristics or else an instruction manual. Look for energy dissipation figures listed in joules. A joule is approximately 0.7375 foot pounds, and is the number of watts consumed in one second. A good rating is 115 joules, but a higher number is preferable. Some units indicate the maximum spike voltage protection. Sometimes this is in practical terms, such as 5 kilovolts and 5,000 amperes, while some indicate higher numbers. The higher the better.

These figures indicate protection against lightning so they should be sufficient to supply protection against in-home appliances. A well-constructed surge protector should also be equipped with a radio-frequency filter as a means of reducing power-line interference. The frequency range of such a filter is enormous and can extend from 150 kHz to 20 mHz. Alternatively, the decrease in interference is specified in decibels, and the higher the better. Look for a bottom amount of 5 dB.

The speed with which a surge suppressor will respond is important. Usually the response is one billionth of a second, or a nanosecond. A longer lasting strong pulse may make the suppressor inoperative and will cut off power to the computer. Some suppressors are equipped with warning lights to indicate if the unit is on, or if it is working properly.

ADDITIONAL WIRING TECHNIQUES

How to Fish Wire

Fishing wire is the technique used to pull cables behind walls. One of the more difficult jobs in electrical wiring is to install wiring in completed structures. So before starting this time-consuming job, determine just what is to be accomplished. Usually, it is to bring wires from a source of electrical power to some other area where that power can be controlled, such as by a switch, or utilized, as in a receptacle.

Work done behind walls is not easy and requires some planning. First, locate a section of the wall that has an open space behind it. That space will extend from floor to ceiling but is shut off at both those areas. One such place is joint B shown in Figure 7–18.

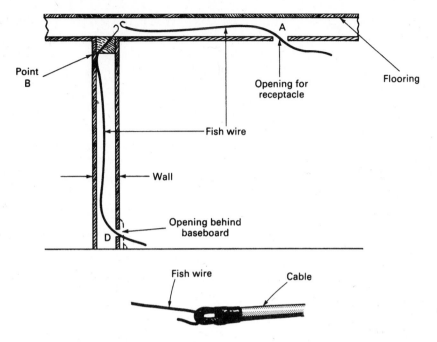

Figure 7–18. Method of fishing wires behind walls.

It will now be necessary to cut an opening at some small distance above the baseboard (D). A typical area would be about 45 inches above the baseboard.

Before cutting into the wall, try to locate a convenient stud. This is a wooden member of the framing of the structure and is a convenient support for an electric box. There are several ways of locating a stud. One is to knock on the wall at various spots. An in-between stud space will sound hollow, but a stud will produce a solid sound. A preferable technique is to use a stud finder. This is an inexpensive compass whose pointer will swing strongly when brought near a stud since studs have nails in them.

Still another way is to drill holes in the wall using a $\frac{1}{8}$-inch drill. Do this near the floor, but first remove the baseboard so that the holes will not show after the baseboard is replaced. Consider point A in the drawing. This would be the ceiling. If this is the location of a light fixture, it represents the power source and this would be where a cable connection would be made.

The next step would be to drill an opening at point B, something that might or might not be possible. Assuming that this is done, use two lengths of fish wire and form loops at each end. Work the wires up and across and try to get the two hook ends to engage. It is not easy and requires patience. Also, it is not a one-man job. It takes two people who have the ability to work together. Fishing wires through a wall is not easy for a professional electrician and much less so for an amateur. If at all possible, get a structural plan of the structure. This will show, at least, where there might possibly be some interference.

Boring a channel for a cable requires a long bit and a wood drill. An opening can be obtained at joint B in Figure 7–18 by using such equipment and drilling downward. This is easier than trying to drill upward. Remove the baseboard of the flooring above joint B and drill downward as indicated in Figure 7–19.

How to avoid drilling.

There are two ways of avoiding drilling. One technique is to use on the surface wiring. This is the easiest method, but it has the disadvantage of showing a channel that covers the cable. Another method is to cut a groove in the wallboard, sheetrock, or plaster that forms the wall. Hold the wire cable in place with staples, taking advantage of the studs for support. After the cable is in position and all electrical connections have been made and tested, plaster over the grooves.

To avoid the need for refinishing the grooves, remove the baseboard and make a channel for holding the cable. After the cable has been installed and tested, replace the baseboard.

How to conceal the work.

Cutting into a wall is always a problem. If the wall is painted, the section of wall that has been removed can be replastered and then painted once again. However, it is not easy to match paints since paint, although it carries the same identification number on the can, may differ slightly from batch to batch. Also, old paint on a wall gradually oxidizes and changes color. The solution may be to paint the entire wall.

Working with wallpaper-covered walls may be a bit easier. When having a room papered, always keep scraps carefully wrapped with some identification on

Figure 7–19. Method of drilling downward to supply access hole for wire. The straight line below the bit indicates an alternative approach.

the outside of the package. Draw a template using a light pencil line (do not use ink or a marking pen), and then use a sharp, preferably new single-edge blade to make the cut. Do not cut freehand; use a ruler.

Make two vertical cuts and one horizontal. The paper will then be in the form of a flap that can be lifted. If the paper tends to stick to the wall, use water as a paste solvent. If this does not work, use a commercial solvent, which can be obtained in a wallpaper store.

After the electrical wiring has been installed and tested, fill the opening with a mixture of plaster and spackle. By itself, plaster dries too quickly. A small amount of spackle delays drying time and supplies a smoother finish. When the plaster is thoroughly dry, paste the wallpaper flap back into position. If the cuts were sharp, then the joined edges of the paper should be barely noticeable.

Fish wire and drop chains.   Fish wire is not only used to pull wire behind walls but is also helpful in getting a wire cable through conduit. The amount of effort required depends on the inside diameter of the conduit, the smoothness of its inner walls, and the number of cables the conduit is expected to carry.

Fish wire is made of tempered steel, has a diameter of about $\frac{1}{4}$ inch, and is available in various lengths. It is rather stiff, but it can be pushed or pulled through elbows of conduit.

How to Connect a Ceiling Fan[10]

There are various ways of connecting a ceiling fan depending on the type of control to be used. The control is a switch, and it can be a pull-chain type that is an integral part of the fan, or it can be an external, wall-mounted SPST switch. The fan may also be equipped with a light, which can also be wall switch or fan switch controlled.

A fan can be wired in three different ways. The wiring of a ceiling fan is very simple and is only a matter of connecting the fan wires to the power cable in a ceiling electric box. Because of its weight and the need for attaching the fan to the box, the physical setup is much more difficult than the electric wiring.

If the fan is to be operated by a pull chain, this means that the operating switch is built into the fan and that no wall control will be used. Examine the wires that extend from the tubular portion. There will be four wires color coded green (ground), white (neutral), black (hot), and blue (hot). The blue line is tapped off the black lead since two voltages will be needed, one for the fan motor and the other for the fan light. The ground lead is connected to a machine screw on the outside tubular portion of the fan assembly. See Figure 7–20.

[10](NEC: Article 680–6; page 767)

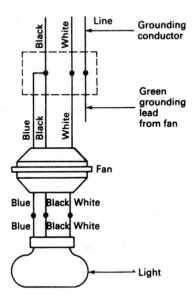

Figure 7–20. Fan controlled by pull-chain switch. Light is controlled by a chain switch on the light.

Another possible arrangement is with a wall control for the light. This circuit (Figure 7–21) shows that the black and blue wires are connected to a wall-mounted switch. The switch should be such that the fan and fan light can be operated separately or simultaneously.

The third circuit is one in which wall controls for the fan and light are used (Figure 7–22). The light assembly is usually supplied in kit form. However, all

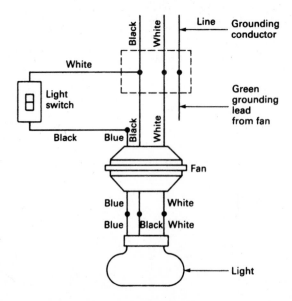

Figure 7–21. Light controlled by wall switch. Fan controlled by a pull chain.

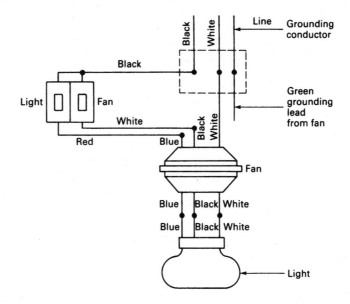

Figure 7–22. Fan and light controlled by independent wall switches.

that is required is to connect the light-control wires to the fan wires, being careful to follow the color coding of the wires, that is, black to black, and white to white.

Electric box for ceiling fan installation.[11] Because of its weight, the electric box used in the ceiling for suspending the fan cannot be the ordinary type commonly used for wall-mounted receptacles and switches. The box that is used should be marked as suitable for fan support and should carry an identifying label to that effect. The electric box for a fan is fastened by a bolt holding it to a telescoping brace. This brace is made specifically for supporting the fan's electric box. If an ordinary hanger bar is used, replace it with the new brace. Municipal codes may require the use of an electric box and telescoping brace made specifically for fan support.

How to Upgrade Receptacles

Electrical devices are now equipped with three-prong plugs. The third prong, easily recognized since it is cylindrical, is a ground. The other two are for connections to a hot lead and to a neutral wire (Figure 7–23). The three-prong plug is part of a three-wire cable. The ground lead of this cable is connected to the metal

[11](*NEC: Article 370–27c; page 309*)

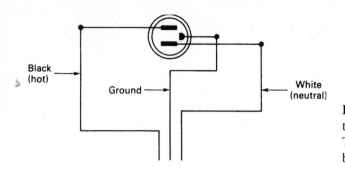

Figure 7–23. Connections to a three-prong polarized receptacle. The three wires lead to a three-wire branch power line.

frame of electrical equipment such as a refrigerator, toaster, or broiler. This grounding is a safety measure and is for the user's protection in the event the metal frame shorts to a hot lead.

To remove an old receptacle, turn off power to the receptacle by removing its fuse from the service box or set its circuit breaker to its off position. To make absolutely sure that the receptacle being removed is not powered, test it with a neon tester. The neon bulb should not light.

For an electric box in the middle of the run, there will be an input and also an output power cable. After the wires have been disconnected from the old receptacle, connect the black lead of the input cable to the brass screw of the upper receptacle, corresponding to the short slot. Connect the white lead of this cable to the silver screw, corresponding to the longer slot. Repeat this wiring procedure using the lower receptacle and the wires of the output cable. Check to make sure that there is a metal connector from the upper silver screw to the lower one and that there is a similar metal connector from the upper brass screw to the lower one. These metal strips supply power from the upper receptacle to the lower.

Note that there are two ground wires. One of these connects to a grounding screw on the lower-left side of the receptacle. The screw head may be color coded green. One of the grounding wires is supplied by the input cable. It should be connected to the grounding screw, as well as the grounding wire of the output cable. The ground wire of the output cable should also be connected to a ground clip of the electric box.

Use a wire nut to cover any wires that are not connected. Push the new receptacle into the electric box and fasten it in place with a pair of machine screws. Make sure the receptacle is centered and that no side screws can touch the metal box. Replace the cover plate. Since this receptacle can accept three-prong plugs, an adapter will not be required.

Chapter 8

Transformers, Capacitors, Inductors, and Relays

TRANSFORMERS[1]

A transformer is probably one of the most efficient of all electrical components, with efficiency values ranging in excess of 90%. A transformer has no moving parts, is made of one or more coils of wire, and depends on the transfer of electrical energy from one coil to another via one or more magnetic fields.

Transformer Categories

Transformers can be categorized in a number of ways: by function, by turns ratio, by power-handling capability, by whether or not they have a core, and by frequency.

Transformer Coils (Inductors)

A transformer may be equipped with one or more coils of wire; if there is more than one, there is no physical connection between them. Transformers can be used as voltage or current step-up or step-down devices.

Figure 8–1 shows a transformer step-down type. It has two windings, one of

[1](NEC®: Article 450; page 471)

225

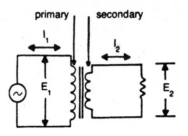

Figure 8–1.   Voltage step-down transformer.

which is identified as the primary and the other as the secondary. The parallel vertical lines represent the transformer core. The core is made of thin sheets of iron or silicon steel, formed in the shape of the letters E and I. The primary and secondary coils are wound of insulated wire around the core, but there is no contact between the wire and the core. After assembly, the unit is sprayed with a fixative to hold the coils in place.

The primary winding is connected to an ac voltage source, usually by a length of wire cord ending in a suitable plug. The alternating current surging back and forth is identified as I_1 and the line voltage across the primary as E_1. In some instances a single-pole, single-throw switch is put in series with the primary winding so as to turn the voltage on and off. In some cases, also, a fuse is put in series with the primary.

The secondary winding of this transformer has fewer turns than the primary. This is done deliberately to emphasize the fact that the transformer is a step-down type; that is, the voltage across the secondary is less than that across the primary. The voltage across the secondary is E_2 and the secondary current is I_2. The resistor across the secondary, shown as a jagged line, represents the load. This load could be a light bulb of some kind or a heating element, or the transformer could be part of a tool or the input voltage to a motor.

Magnetic Induction

Every wire carrying an alternating current, whether the wire is straight or coiled, is accompanied by a varying invisible magnetic field. This magnetic field grows from zero to some maximum value depending on the strength of the varying current through the coil. The magnetic field surrounds both the primary winding and the secondary winding. In so doing, the magnetic field induces a voltage across the secondary. This voltage is ac and has the same frequency as that appearing across the primary.

The strength of the original magnetic field depends on the current flowing through the primary, on the number of turns of primary wire, on the way the primary is wound, on the number of layers of wire, on the type of metal used as the core, and on the inductance of the primary. The greater the inductance the stronger the magnetic field, and this is the reason for having an iron or steel core. It increases the inductance.

Transformer Types

Open or closed transformers.   The transformer may be enclosed in a metal casing, in which case it is referred to as a closed transformer. Alternatively, it may not have a metal enclosure and so is categorized as an open type. Ordinarily, the metal casing is grounded. The basic purpose of the casing is to confine the magnetic fields to the immediate vicinity of the transformer and to prevent them from inducing unwanted voltages across other conductors. Since the transformer may become very warm during the time it is functioning, the casing can also work as a heat sink; that is, it can radiate heat away from the transformer.

Step-up transformer.   If the secondary winding of the transformer has more turns than the primary, as shown in Figure 8–2, the unit is known as a step-up type. In this case the voltage across the secondary is higher than that across the primary. The transformer is also a current step-down type.

One-to-one transformer.   Figure 8–3 shows a transformer having an equal number of primary and secondary turns. It is also sometimes referred to as an isolation transformer; is intended to be used when it is necessary to isolate the load from the primary source voltage.

Distribution transformers.   These are pole-mounted types having an input of 2400 volts and an output of 240 volts, but they can also have their secondary windings arranged for 120-volt output. These are single-phase transformers.

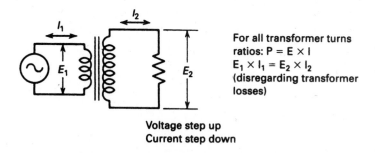

For all transformer turns
ratios: $P = E \times I$
$E_1 \times I_1 = E_2 \times I_2$
(disregarding transformer
losses)

Voltage step up
Current step down

Figure 8–2.   Voltage step-up transformer.

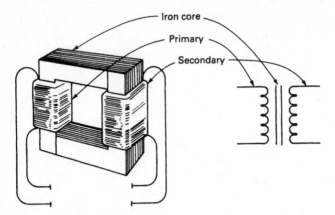

Figure 8–3. One-to-one transformer has equal number of primary and secondary turns.

Transformer Losses

As in all other electrical devices, transformers have power losses that reduce transformer efficiency (Figure 8–4). One of the main reasons for power loss is the heat generated by the transformer. This is due to the flow of current through the transformer coils and can be substantial when large amounts of current are involved, as, for example, in pole transformers used by electric utilities. To help move the heat from the laminations and the transformer coils, such transformers are filled with a special type of oil known as transformer oil. This oil has good heat conductivity, which delivers the heat to the transformer's metal casing where it is radiated to the outside air. The casings are designed to present a large surface area for maximum heat radiation ability.

Another cause of power loss is due to the core. Since it is made of metal, the varying magnetic field around the primary and the secondary induces circulating currents in the core. These currents, known as *eddy currents,* perform no useful

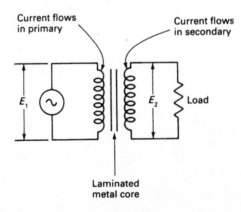

Figure 8–4. Secondary current is reduced by effect of transformer losses.

work and represent an energy loss in the form of heat. Eddy currents are minimized by building the core of thin metal stampings known as *laminations.*

The laminations are made of heat-treated, grain-oriented silicon steel. The laminations are covered with a special varnish used to insulate adjacent laminations from each other.

Transformer Power Relationships

The power delivered to the primary winding of a transformer can be written as

$$P = E_1 \times I_1$$

and the power supplied by the transformer to the secondary winding is

$$P = E_2 \times I_2$$

where E_1 is the voltage across the primary, I_1 is the primary current, E_2 is the voltage across the secondary, and I_2 is the secondary current. If we disregard the losses of the transformer, then (Figure 8–5)

$$E_1 \times I_1 = E_2 \times I_2$$

With this equation it is possible to calculate any one of these values if the other three are supplied. Thus,

$$E_1 = \frac{E_2 \times I_2}{I_1}$$

$$I_1 = \frac{E_2 \times I_2}{E_1}$$

To calculate either the secondary voltage or current:

$$E_2 = \frac{E_1 \times I_1}{I_2}$$

$$I_2 = \frac{E_1 \times I_1}{E_2}$$

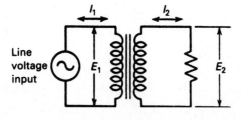

Line voltage input

Figure 8–5. Primary current × primary voltage = secondary current × secondary voltage, assuming no transformer losses.

The product of E and I is the power in watts. To state that $P_2 = P_1$ means that the losses in the transformer are ignored; that is, 100% efficiency is assumed.

Percentage transformer efficiency.
If the efficiency factor of a transformer is high, it may be possible to ignore it, depending on the accuracy required by the problem. The efficiency factor is determined by the power loss. But power is $P = E \times I$. The voltage, E_t is determined by the ratio of secondary and primary turns and is fixed. This means the loss is evidenced by a reduction of current flow.

As an example, consider a 1:1 (one to one) isolation transformer. Because the primary and secondary windings have the same number of turns, the primary and secondary voltages are equal. If the primary voltage is 120 volts, the secondary voltage is also 120. The secondary is 120 volts whether it is shunted by a load or not, and it is 120 volts whether the load is a heavy one (large current flow) or light (small current flow). When the load is heavy, the primary power increases and so does the secondary power. The voltages across the primary and secondary remain unchanged. With a heavy load, a correspondingly heavier current flows in the primary and secondary. If the secondary load is removed, the primary current drops practically to zero. But when the secondary load is removed, the voltage across the secondary remains the same. With no current flowing in the secondary winding (Figure 8–6)

$$P = E \times I$$
$$P = 120 \times 0 = 0$$

Turns ratio for voltage.
The voltage and current available from the secondary winding of a transformer depend on the number of turns of wire of the secondary coil of the transformer compared to the number of turns on the primary. This comparison is called the *turns ratio* and is determined by dividing the secondary turns by the primary turns or by dividing the primary turns by the secondary turns.

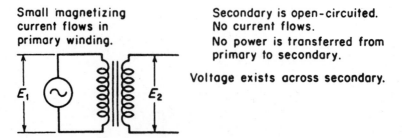

Small magnetizing current flows in primary winding.

Secondary is open-circuited. No current flows.

No power is transferred from primary to secondary.

Voltage exists across secondary.

E_1 E_2

Figure 8–6. When secondary winding is open circuited, voltage appears across the winding, but there is no current flow. Power developed across secondary is zero.

$$T_r = \frac{N_2}{N_1}$$

T_r is the turns ratio, N_2 is the number of secondary turns, and N_1 the number of primary turns. If, for example, a transformer has 30,000 secondary turns and the primary has 10,000 turns,

$$T_r = \frac{30,000}{10,000} = \frac{3}{1}$$

The turns ratio is 3 to 1. There are three secondary turns for every turn on the primary.

In an alternative example there are more primary than secondary turns. For example, a transformer having 600 primary turns and 200 secondary turns is

$$T_r = \frac{200}{600} = \frac{1}{3}$$

There is one secondary turn for every three turns on the primary.

Relationship of transformer voltage to the turns ratio. Knowing the turns ratio of a transformer and the primary voltage makes it possible to calculate the secondary voltage. The voltage across the secondary (E_2) is equal to the primary voltage (E_1) multiplied by the turns ratio.

$$E_2 = E_1 \times T_r$$

From this formula it is possible to derive two more:

$$E_1 = \frac{E_2}{T_r}$$

$$T_r = \frac{E_2}{E_1}$$

E_2 is sometimes written as E_s and E_1 may be found as E_p.

The ratio of the secondary voltage divided by the primary voltage is equal to the turns ratio. This statement can more conveniently be expressed in terms of a formula:

$$\frac{E_2}{E_1} = \frac{N_2}{N_1}$$

The short horizontal rule beneath E_2 and N_2 indicates division. This division symbol can also be written as ":" and so the formula becomes

$$E_2 : E_1 = N_2 : N_1$$

If we transpose E_2 and E_1, the equality will remain unchanged, provided we do the same to N_2 and N_1. Thus, the formula becomes

$$E_1 : E_2 = N_1 : N_2$$

There are four terms in this formula. If any of the three are known, the fourth can be calculated.

$$E_1 = \frac{E_2 \times N_1}{N_2}$$

$$E_2 = \frac{E_1 \times N_2}{N_1}$$

$$N_1 = \frac{E_1 \times N_2}{E_2}$$

$$N_2 = \frac{E_2 \times N_1}{E_1}$$

A comparison of the primary and secondary currents to the turns ratio shows that the current ratio is the inverse of the voltage ratio. This is just another way of saying that if you step up the voltage you step down the current. Conversely, if you step up the current, the voltage comes down.

Current ratio

$$\frac{I_1}{I_2} = \frac{N_2}{N_1}$$

Voltage ratio

$$\frac{E_2}{E_1} = \frac{N_2}{N_1}$$

Just as it was possible to show the relationship between the turns ratio and the primary and secondary voltages, it is also possible to show the relationship between the turns ratio and the primary and secondary currents.

$$I_1 = \frac{I_2 \times N_2}{N_1}$$

$$I_2 = \frac{I_1 \times N_1}{N_2}$$

$$N_1 = \frac{I_2 \times N_2}{I_1}$$

$$N_2 = \frac{I_1 \times N_1}{I_2}$$

Transformers Location[2]

Transformers must normally be accessible for inspection (NEC 450-13, page 479), but transformers having a high voltage or a high kilovolt-ampere rating, when installed indoors, must be in a transformer room or vault. The construction of the vault is covered in the code.

Overcurrent protection for transformers is based on their rated current and not on their load. NEC also contains rules for the installation and protection of transformers up to 600 volts.

CAPACITORS[3]

A capacitor consists of any pair of adjacent conductors that can store an electric charge. In its simplest form a capacitor is a pair of metal plates that may be adjacent. A capacitor may be deliberately constructed for certain of its properties, or it may be accidental. A line cord, for example, exhibits capacitance between its wires.

While an inductance can store energy in an electromagnetic field, a capacitor stores energy in an electrostatic field. The capacitance of a capacitor is a measure of its ability to store an electric charge.

Capacitor Construction

Basically, a capacitor consists of a pair of metal plates or conductors separated by a dielectric (Figure 8–7). The dielectric can be air, glass, mica, or some other substance. The amount of capacitance depends on the area of the plates or conductors and the type of dielectric used. The greater the plate area, the larger the amount of capacitance. Capacitance also depends on the type of dielectric that is used and is known as its *dielectric constant*. A vacuum is taken as the reference and has a constant of 1. Glass, depending on its type, has a dielectric constant of 5 to 10. Thus, if glass is substituted for a vacuum in a capacitor, the amount of capacitance will increase by a factor of 5 to 10.

Another factor that determines the amount of capacitance is the distance between the plates or conductors of the capacitor. The closer the plates are, the greater the capacitance.

[2]*(NEC: Article 450–13; page 479)*
[3]*(NEC: Article 460; page 486)*

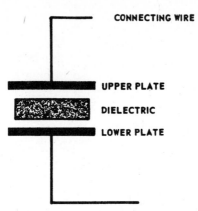

CONNECTING WIRE

UPPER PLATE

DIELECTRIC

LOWER PLATE

Figure 8–7. Basic capacitor construction.

Voltage Rating of a Capacitor[4]

In selecting or substituting a capacitor for use in a particular electrical application, certain considerations are involved. One is the amount of capacitance, and another is the working voltage, sometimes expressed as dc working voltage, WV or DCWV. If the working voltage is exceeded, the dielectric will break down and arcing will occur between the plates. The working voltage of the capacitor is the maximum voltage that can be applied without danger of arc-over. The working voltage depends on the kind of material used as the dielectric and its thickness.

Capacitor Circuit Conductors

When a voltage is first applied to a discharged capacitor, there will be an inrush of charging current. So the conductors connected to the capacitor must have an ampacity of not less than 135% of the rated current of the capacitor.

Capacitors in Parallel

Capacitors can be connected in parallel to supply a greater amount of capacitance. When all the capacitors (Figure 8–8) are in the same capacitance units, all that is necessary is to add the capacitance of the individual units. If dissimilar capacitance units are used, it will be necessary to convert. Thus, if 2 μF, 6 μF, and 2000 pF are to be wired in parallel, it will be necessary to convert the 2000-pF unit to its equivalent in microfarads. The formula for capacitors in parallel is

$$C_t = C_1 + C_2 + C_3 \ldots$$

[4]*(NEC: Article 460–12; page 488)*

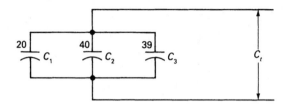

Figure 8–8. Capacitors in parallel.

There are numerous capacitor types, and while dissimilar capacitors can be wired in parallel, the more desirable arrangement is to connect those that are identical. All the capacitors should have the same working-voltage characteristics. When capacitors are connected in parallel, the total capacitance is always larger than that of the highest capacitance value of any individual capacitor.

Capacitors in Series

Capacitors can be wired in series as indicated in Figure 8–9. The sum of the capacitor voltages will be equal to the source voltage. The total capacitance of the series group is

$$C_t = \frac{C_1 \times C_2}{C_1 + C_2}$$

All the capacitors must be in identical units. When capacitors are connected in series, the total capacitance will always be less than that of the smallest-value capacitor in the circuit.

Capacitor Current and Voltage Phase Angle

An uncharged capacitor will draw a large current when a dc voltage is first applied. As current flows into the capacitor, the voltage across it will rise. After the capacitor has received a full charge, the voltage across the capacitor will be equal to the applied voltage, and no further current will move into the capacitor.

If, instead of dc, an ac sine-wave voltage is applied, the current input to the capacitor is maximum, and the charge voltage of the capacitor will start to rise. The

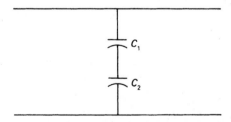

Figure 8–9. Capacitors in series.

current is zero when the voltage across the capacitor is maximum. In a perfect capacitor, that is, one that contains no resistance, the current will lead the applied voltage. Phrased in a different way, the voltage can be said to lag the current.

Capacitive Reactance

The reactance of a capacitor, stated in ohms, is its opposition to the flow of an alternating current. The symbol for this reactance is written X_c. Its value is inversely proportional to the applied frequency and the amount of capacitance expressed in farads. In terms of a formula,

$$X_c = \frac{1}{2\pi f C}$$

π is the Greek letter pi and has a fixed value of 3.14159, and f is the frequency in hertz. f is most commonly 60 hertz, although other frequencies (less often used) in electrical work are 25 and 50 hertz. Note that as the frequency or the capacitance increases the reactance decreases.

Utilization of Capacitors in Electrical Work

The voltage across a resistor connected to an ac receptacle is in phase with the current flowing through it. If the resistor is put in series with a capacitor, the voltage across the capacitor and that across the resistor will be out of phase by a maximum of 90°. Thus, these two components can produce the equivalent of a two-phase setup and so are suitable for two-phase operation.

This type of circuitry is useful for three basic types of ac motors: capacitor start, split capacitor, and two-value capacitor. Capacitors are commonly used to supply a substantial starting torque, since their use supplies the equivalent of a rotating magnetic field.

INDUCTORS

An inductor is a coil of wire and as such has an electrical property known as *inductance*.

There are two basic types of inductors, air core and iron core. Iron-core units are available in two forms, fixed core and variable core. The iron core increases the inductance of the coil, and if it is a movable core, it is able to vary the overall inductance.

Phase Relationships of an Inductor

When a coil is connected to an ac source, the current lags the voltage by a maximum of 90°. This phase difference can be lowered depending on the amount of resistance inherent in the wire forming the coil.

Induced Electromotive Force

The passage of a current through a conductor, even if that conductor is a straight length of wire, is accompanied by a magnetic field. If the current is dc, the magnetic field remains steady once the current reaches its operating value. If the current is ac, the magnetic field varies in step with the frequency of the applied voltage.

When a varying current, such as that supplied by an ac voltage, flows through a conductor, its varying magnetic field cuts across that conductor and induces a voltage across it, referred to as a self-induced electromotive force (emf). This self-induced emf can be very small for a straight wire, but is increased considerably when the wire is wound into the form of a coil (also known as an *inductor*).

The amount of inductance is also affected by the way in which the coil is wound. It is lower if the turns of wire of the coil are spaced apart, but higher if they are wound in layers, if the number of turns is increased, or if an iron core is used.

RELAYS

When a coil is connected to a dc voltage source, the current flowing through the coil is accompanied by a magnetic field. The effect is to make the coil into an electromagnet. There is no visible physical change in the coil, but it now behaves in the same manner as a permanent magnet. The coil is referred to as a temporary magnet because it loses its magnetism when the flow of current is stopped.

Figure 8–10 shows the structure of a relay. Like so many other electrical components, the relay is available in a large number of sizes, styles, and capabilities. Figure 8–11 shows a relay used for the control of a pair of motors, identified as 1 and 2.

The source of dc voltage is $B1$ and it is connected to a resistor R, which controls the amount of current flowing through the wires of coil, C. A small section of metal, such as steel (A) is hinged at one end and is connected to a spring identified as S. In the absence of current flow through the coil, the spring holds the metal strip, called an *armature,* against the right-hand contact of a single-pole, double-throw switch (SPDT). This completes the circuit for motor M2, which now receives its operating current from battery $B2$.

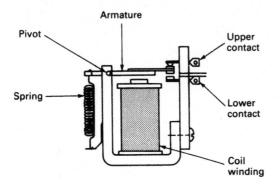

Figure 8–10. Relay structure.

The relay coil can be energized by reducing the resistance of *R*. In some circuits a switch is inserted in series with *R,* and closing the switch permits current flow through coil *C,* energizing it. The resulting magnetic field attracts the armature, closing the circuit to motor M1. With this circuit, a relay can be used to operate either one of two motors.

This is just one of many possible applications for relays. Unlike mechanical switches, they do not need manual closure and so can be used to close a circuit at a distance. They can also be made to operate automatically with the help of an

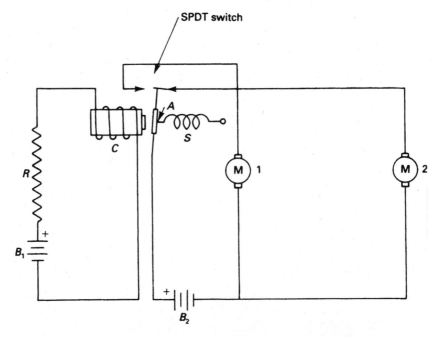

Figure 8–11. Relay circuit for motor control.

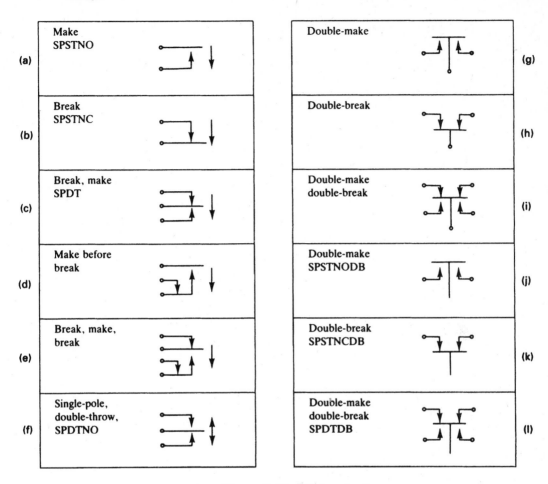

Figure 8–12. Relay circuits.

electric timer. Relays can be used in environmental conditions that are hazardous. They can turn high voltages on or off when these voltages are necessary for operating equipment. They can be used to protect a motor against phase failure or phase reversal. This latter condition can cause a motor to turn in an opposite direction, a condition that could be hazardous in the case of elevators, hoists, and cranes. Relays can be used to control the upward or downward movement of elevators. A relay can be used to operate a large number of switches, such as those shown in Figure 8–12. These can be listed as:

1. SPSTNO—single-pole single-throw normally open (make)
2. SPSTNC—single-pole single-throw normally closed (break)
3. SPDT—single-pole double-throw (break, make)

4. SPST—single-pole single-throw (make before break)
5. SPDT—single-pole double-throw (break, make, break)
6. SPDTNO—single-pole double-throw, normally open
7. DPST—double-pole, single-throw (double-make)
8. DPST—double-pole, single-throw (double-break)
9. DPDT—double-pole, double-throw (double-make, double-break)
10. SPSTNODB—single-pole, single-throw, normally open, double-break (double-make)
11. SPSTNCDB—single-pole, single throw, normally closed, double-break
12. SPDTDB—single-pole, double-throw, double-break

A relay switch can be normally open; that is, in the absence of a relay current, the relay coil is not energized, and the relay contacts do not touch. A switch that is normally open (NO) is a make type; that is, the switch contacts must meet or make. A switch that is normally closed (NC) in the absence of relay current is a break type; that is, the switch contacts must open or break.

Chapter 9

Indoor and Outdoor Lighting[1]

Just as many homes are inadequately wired, so too do many homes use poor lighting. However, there is an important difference. Poor lighting does not always mean inadequate lighting, for it can also indicate excessive lighting or lighting that is too concentrated or not soft enough. It is possible to have a large room with a strong light at one end with the rest of the room practically in darkness.

Basements and attics, particularly in older homes and warehouse space used by business, are sometimes so poorly lighted that they are accident traps. All this is unfortunate because it is unnecessary. Outdoor walkways can also be hazardous, which is also unfortunate since such lighting is easily installed using low voltage.

Good lighting is readily obtainable. Also, compared to some power-hungry appliances, good lighting can be a best electrical buy. A toaster will gulp 1200 watts, but a 100-watt electric-light bulb uses 10% of this amount. But even with light bulbs, careful selection and use can reduce operating costs.

[1](NEC®: Article 220–3(b); page 69)

LIGHTING FACTORS[2]

Light and Vision

Electric light is visible in two ways, directly and by reflection. Looking into the lighted bulb of a flashlight is an example of direct light. But for the most part, the light that we use is reflected. Light shines on some object and is bounced from it. Obviously, then, one aspect of good lighting is to make sure that objects reflect enough light to become visible.

The Ability to See

Four factors determine seeing ability:

1. *Size:* The larger an object is, the more light it can reflect. A larger reflecting surface means greater visibility.
2. *Contrast:* An object that is uniformly dark in color will not be seen as easily as one made of contrasting colors.
3. *Reflectivity:* Some colors reflect light better than others. A white wall will reflect light, if any light is available, better than one that is black.
4. *Time:* It takes time for eyes to adapt to light. In a darkened room the eyes adjust to receive the maximum reflected light. But going out into sunshine means the eyes adapt once again. It is helpful when going from a well-lighted room to one that is in darkness to close the eyes for a few moments to give them time to adapt to seeing in greatly reduced light.

Light Comfort

Getting the right amount of light is just one step since there is also the problem of light comfort. To make sure light is comfortable, it must be well distributed and free from glare, with no bright spots in the room and an absence of deep shadows.

When selecting a particular wall paint, wallpaper, or wall fabric, examine it in two ways. Get as large a sample as possible and look at it under working lighting conditions. A sample of wallpaper or fabric will have one color in a showroom and possibly another in the room being considered because the lighting conditions are different.

These cautions also apply to paint. Don't start by painting an entire room.

[2]*(NEC: Article 410; page 358)*

Try some on a section of wall and then consider it under two conditions: the first with natural light from the windows with shades and drapes in their customary positions, and the second with artificial light supplied by floor, wall, and ceiling lights. Do this only after the paint has completely dried. The color of the paint and its reflectivity, that is, whether it is a gloss type such as enamel or a flat paint such as most latex types, will make a big difference.

Effect of Color

White in a room supplies the greatest amount of reflectivity. A room with white ceilings and white or near-white walls makes it possible to use lights with a lower wattage rating. This is also true if much of the furniture is either white or close to it in color. But with this color arrangement, if unnecessarily higher wattage bulbs are used, there may be too much glare.

The ceiling and/or wall colors used in a room depend on the amount of available daylight. Strong colors can be used if the daylight is substantial. Such colors, though, can make a room look smaller. Use light tints to make it seem larger.

Approximate Reflectance Values

Table 9–1 supplies the approximate percentage of reflection of light by various colors. No matter what the room size may be, the safest course to follow is to make a selection of ceiling and wall colors in about the 30% to 80% reflectance range. The usual routine is to have the ceilings painted white while floors are made much darker. A room will appear brighter if the floor is covered with a white or close to white rug.

A white ceiling acts as a large reflecting surface (Figure 9–1) and also diffuses the light making it softer. It also helps spread light over an entire room. There are interior-design arrangements in which the ceiling is papered or covered with fabric with both having a nonwhite pattern. These reduce the amount of reflected light. While this problem can be overcome by adding more lights or higher-powered bulbs, such rooms can look gloomy if these additions are not made.

Dimmer Control Effect

With incandescent lighting, increasing the dimming action that is, by reducing the overall light level, has the apparent effect of changing the light of the bulbs from white light to yellow. Yellowish light can be soft and pleasing. Whether yellow light is desirable or not is a personal matter.

TABLE 9–1   COLOR REFLECTANCE VALUES

Color	Approximate Percent Reflection
Whites	
Dull or flat white	75–90
Light tints	
Cream or eggshell	79
Ivory	75
Pale pink and pale yellow	75–80
Light green, light blue, light orchid	70–75
Soft pink and light peach	69
Light beige or pale gray	70
Medium tones	
Apricot	56–62
Pink	64
Tan, yellow-gold	55
Light grays	35–50
Medium turquoise	44
Medium light blue	42
Yellow-green	45
Old gold and pumpkin	34
Rose	29
Deep tones	
Cocoa brown and mauve	24
Medium green and medium blue	21
Medium gray	20
Unsuitably dark colors	
Dark brown and dark gray	10–15
Olive green	12
Dark blue, blue-green	5–10
Forest green	7
Natural wood tones	
Birch and beech	35–50
Light maple	25–35
Light oak	25–35
Dark oak and cherry	10–15
Black walnut and mahogany	5–15

Ambient Light

Ambient light consists of light reflected from walls, ceiling, furniture, and lamp-shades. It also includes light coming in through windows or any other access to outside light.

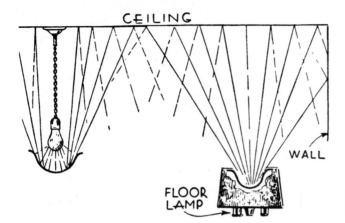

Figure 9–1. Ceiling acts as light reflector.

Sources of Light

There are two sources of light: natural and artificial. Sunlight is natural light. Light from incandescent, fluorescent, halogen bulbs, candles, or a fireplace is considered artificial.

Inverse Square Law

The illumination that falls on a surface varies inversely as the square of the distance from the light source. As an example, assume a 100-watt light bulb supplies satisfactory illumination on an object that is 5 feet away. If the object is moved to a greater distance, such as 10 feet, the original distance has been multiplied by 2. But the illumination varies inversely as the square of the distance. $2^2 = 4$. Thus, at a distance of 10 feet we will need four 100-watt bulbs to illuminate the same surface area to the same extent.

Dirt and Illumination

After being in use for some time, bulbs get dirty. The dirt forms an opaque surface coating, and while it does not cut out the light completely, it does reduce it. However, if a 75-watt bulb becomes coated with dirt, the cost is still for 75 watts of electrical power. A dirty 75-watt bulb might supply only the equivalent of a clean 50-watt bulb. The first step to getting more illumination is to clean the bulbs.

Use a dry cloth. Moisture on the surface of a light bulb will cause it to crack as soon as the bulb gets hot. There is no harm in removing bulbs and washing them, provided the bulb is completely dry before putting it back in its socket.

TERMINOLOGY

The light supplied by a source, such as an electric light bulb, is referred to as *luminous flux.*

The unit of luminous flux is the *lumen,* which indicates the total amount of light emitted by a light source. A lumen is the quantity of light covering a surface area of 1 square foot when the source light and the surface area are separated by a distance of 1 foot.

Luminance is the degree of brightness and is measured in lumens.

At one time the amount of light supplied by a source was indicated in terms of candles or candlepower, but these terms are now obsolete. In their place we now have the *candela* (cd), and this is the standard unit of light source intensity. The candela is used in scientific analyses of light, but it has little application in practical electricity.

The candela is the luminous intensity of 0.0167 square centimeter of a black-body radiator at a temperature of 2046 kelvins, which is the temperature at which platinum in a gaseous state begins to solidify. A 40-watt incandescent bulb operating from a line voltage of 120 volts has a luminous intensity of approximately 3000 candela.

LIGHT BULBS

Wattage Rating of a Bulb

The wattage rating of a bulb is indicated in two ways. It will be on the packing box holding the bulb and should also be on the glass end of the bulb. This wattage rating is sometimes difficult to see on the bulb. Blowing on it or holding it up to a strong light may help.

Removing a Light Bulb

Incandescent light bulbs get hot fast. The heat is incidental, that is, not wanted, but it is there just the same. To remove a bulb that has been on for even a short time, switch it off and give it a chance to cool or else use gloves or a bit of scrap cloth when removing it.

Bulb Burnout

Incandescent bulbs have a positive temperature coefficient of resistance. Their resistance is lowest at the start; hence bulb current is highest. If the ac operating voltage is at its peak at this time, the combination of maximum voltage and current can lead to a burnout.

TYPES OF BULBS

Bulbs can be categorized by their electrical design. A basic listing of light bulbs would include the following:

Incandescent (tungsten), clear and frosted
Low voltage (12 or 24 volts)
Halogen (metal halide)
Low-pressure sodium vapor
High-pressure sodium vapor
Mercury vapor
High-intensity discharge (HID)
Fluorescent (rapid start, starter type, instant start, circular, straight, warm, cool)
E lamps

These lights can be subdivided into those designated for a specific use, whether intended for indoor or outdoor use, or by some physical characteristic.

Floodlights	Tinted bulbs
Colored bulbs	White bulbs
Three-way bulbs	Miser (long life)
Spotlights	Chandelier
Mushroom	Earth lights
Well lights	Frosted
Night lights	Frosted reflector
Security lights	Bayonet base
Entrance lights	Miniature screw base
Tier lights	Torpedo
Globe lights	Flame tip (candle)
Spike lights	Complexion (round)
Flashlights	Heat lamps
Decorative bulbs	Silver bowl lamps
Sun lamps	Clear bulbs

Reflector Bulbs

Reflector bulbs have a built-in reflector and are identified by the letters R and PAR. An R bulb puts out about twice as much light as a regular incandescent. A PAR uses a parabolic reflector. It can be used indoors or out and supplies about four times the light than a typical incandescent.

Low-voltage Bulbs

Low-voltage bulbs require a step-down transformer. They are equipped with a built-in reflector with a halogen bulb for accent lighting. The MR-16 is a small size type and is used for recessed and track lighting. The PAR 36 can be used indoors or out and is also a parabolic type.

High-intensity Bulbs

High-intensity bulbs supply much brighter light than incandescent types. They include mercury vapor and high-pressure sodium and are intended for outdoor use.

Frosted and White Finish Bulbs

Frosted and white finish bulbs diffuse the light and help avoid excessive bright spots. For a given wattage rating, they do not yield as much direct light as clear bulbs. They are available as incandescent types, and in other bulb categories as well.

INCANDESCENT BULBS

Incandescent bulbs are sometimes called tungsten bulbs because of the metal used in the construction of its filament. Tungsten was selected for this function since it has a high-melting point and a moderate amount of resistance.

Incandescent bulbs are available in a large variety of physical sizes, shapes, and wattage ratings. Some are designed to flicker, thus simulating the light of a candle. The wattage ratings can vary from as little as 4 watts for a night light to as much as 300-watt floodlights.

Of all the electric bulbs, the incandescent, the most commonly used, is the least efficient, has the smallest useful life, and requires more electrical power for a given amount of light. However, not all incandescent bulbs are alike and some supply more light per wattage input. Thus, low-wattage incandescent bulbs do have a longer life than higher-wattage types, but larger-wattage bulbs are more efficient. Thus, a 100-watt bulb can supply as much light as a pair of 60-watt units, which require a total of 120 watts of input electrical power. However, the 100-watt bulb will have a useful life of about 75% that of the 60 watt bulbs.

Efforts have been made to extend the working life of incandescents. One of these, known as a miser or long-life bulb, achieves this by using a stronger filament. During its working time, the filament of an incandescent bulb gradually vaporizes. Since the miser bulb has a thicker filament, it can last longer, but the thickness means it requires a greater energy input. Since the miser bulb works with a lower wattage input, it produces a smaller light output.

Figure 9–2 shows the structure of an incandescent bulb. The filament is heated by an electric current, but the objective is to obtain light; the heat that is generated can be regarded as an unwanted expenditure of energy.

Incandescent bulbs are vacuum types. The presence of even a small amount of air in the bulb will result in its destruction. During its operating life, the efficiency of the incandescent decreases, and it ultimately loses about 25% of its useful light output due to loss of filament structure by evaporation.

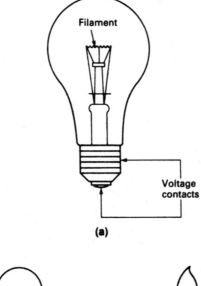

(a)

(b)

(c)

Figure 9–2. Structure of an incandescent bulb (a). Commonly used bulb shape, (b) globular and (c) decorative flame or teardrop. Bulbs can be clear or frosted.

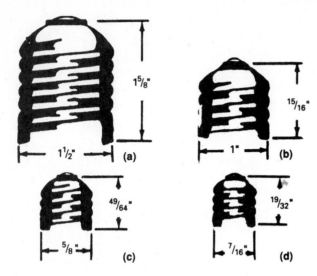

Figure 9–3. Lamp socket sizes. (a) Mogul; (b) medium; (c) intermediate; (d) candelabra.

Socket Sizes

Incandescent bulbs are available with a number of different base sizes: mogul, medium, intermediate, and candelabra. All these are screw types with the bulb base threaded for fitting into corresponding sockets. For insertion, the bulb is rotated clockwise; remove by turning counterclockwise. The medium size (Figure 9–3) is the most commonly used in the home and has a diameter of 1 inch and a length of 15/16 inch. All dimensions are supplied in the drawing (*NEC: Article 410–27; page 365*).

Figure 9–4 shows miniature versions of the incandescent bulb. Some of these have screw bases, but some are also bayonet types. The bayonet is a twist lock. Insert the base in it socket, push the bulb, and then turn in a clockwise direction.

Three-way Incandescent Bulbs[3]

The three-way bulb is a member of the incandescent bulb family. It has two filaments and must be used with a special socket designated as a three-way. Three-way bulbs may be equipped with medium or mogul bases. The bulbs can be rated as 30/70/100 watts, 50/100/150 watts, 50/200/250 watts, and 100/200/300 watts. Table 9–2 lists the sizes and uses for three-way bulbs. The filaments of these

Figure 9–4. Screw and bayonet miniature bulbs

[3](*NEC: Article 410–70; page 374*)

TABLE 9–2 SIZES AND USES FOR THREE-WAY INCANDESCENT BULBS

Socket and Wattage	Description	Where to Use
Medium		
30/70/100	Inside frost or white	Dressing table or dresser lamps, decorative lamps, small pin up lamps
50/100/150	Inside frost or white	End table or small floor and swing-arm lamps
50/100/150	White or indirect bulb with built-in diffusing bowl	End table lamps and floor lamps
50/200/250	White or frosted bulb	End table or small floor and swing-arm lamps, study lamps with diffusing bowls
Mogul (large)		
50/100/150	Inside frost	Small floor and swing-arm lamps
100/200/300	White or frosted bulb	Table and floor lamps

bulbs can be operated separately and can supply different amounts of light, with the operating wattage selected by either a pull chain or a knob mounted on the socket.

If any one of the filaments burns out, you can still use the bulb. In a 50/100/150-watt bulb, if the 50-watt filament burns out, you can still use the bulb as a 100-watt unit. There will be no light when the pull-chain or knob control is set to the 50-watt position.

Clear Incandescent Bulbs[4]

A clear bulb uses a transparent glass structure with the filament easily visible. The chief advantage of this bulb is that it supplies the maximum amount of light for a given wattage rating and so supplies optimum lighting economy. Because they glare, they are often covered with light diffusers. Without this modification, they are used in basements, attics, and garages. Clear bulbs are commonly in globular form, but are also available in torch form for use in chandeliers.

Frosted and White Finish Bulbs

To avoid glare, incandescent bulbs can have either a frosted or white finish. Such bulbs do not supply as many lumens per watt as the clear type.

[4](*NEC: Article 410; page 358*)

Efficiency of Incandescent Bulbs

The efficiency of a light bulb is the total light it emits per watt of electrical power. A small incandescent bulb, such as a 25-watt type, may have an output of 10 lumens per watt. Larger bulbs are more efficient, with a 100-watt bulb supplying about 16 lumens per watt. One 75-watt bulb gives off more light than three 25-watt bulbs. The efficiency of bulbs drops off as they are used, since the bulbs may blacken internally or they may become coated with dirt.

FLUORESCENT LIGHTS[5]

A fluorescent bulb is not just an improved version of an incandescent type; it is a complete departure in design. Its parts are illustrated in Figure 9–5 and consist of a glass tube, the lamp holder, a ballast unit, and, in older units, a starter. Unlike incandescent bulbs, the fluorescent can come equipped with a lamp holder and so may not require a separate fixture.

This bulb uses a pair of electrodes referred to as *cathodes*. When connected to a voltage source, an electric arc is generated between these cathodes, with this arc passing through a mixture of argon gas and mercury vapor. As a consequence,

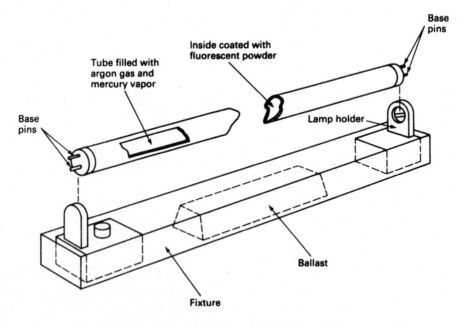

Figure 9–5.   Structure of a fluorescent tube.

[5](*NEC: Article 410–73e; page 374*).

this action releases invisible ultraviolet energy. This energy impinges against the inside wall of the fluorescent tube, which has been coated with a phosphor material that glows in the presence of this energy. Control of the color of the light is produced by a selection of different phosphors. The colors commonly used include cool white, warm white, and daylight.

Fluorescent lights can be categorized in various ways, and the selection of a particular bulb often means making a choice among these characteristics. One of these is the shape of the bulb; it can consist of a straight tube having a length ranging from about 10 inches to as much as 8 feet.

There are three bulb types: preheat, rapid start, and instant start. The preheat unit is one of the older bulbs and has a starter that heats the cathodes. The newer tubes, the rapid start and the instant start, do not have a starter, but obtain a starting voltage from the ballast. The ballast is a step-up transformer applying a higher-than-line voltage to the cathodes. This voltage boost produces an electric arc through the tube (*NEC: Article 410–76a; page 375*).

There is a difference in the operating characteristics of the rapid-start and instant-start tubes. The rapid start requires a few seconds before the light turns on. This may or may not be accompanied by flickering or flashing. The instant start turns on practically immediately without this light behavior.

The fluorescent has four different styles: tubular, circular, U-shaped, and globular. The first three of these have a pair of pins at each end for connection to the operating voltage. The globular is equipped with a base that is similar to that used by incandescent lights.

Unlike the incandescent bulb, which gets hot to the touch in just a few seconds, the fluorescent has a low operating temperature. It is this operating feature that contributes to the efficiency of this bulb. The ambient temperature in which the fluorescent operates should not drop below 65°F for it will result in working problems. The tube is also sensitive to repeated operation of its on-off switch, which could result in tube damage.

The fixture for the tubular fluorescent has a pair of slots at each end designed to receive the pins at opposite ends of the tube. To mount the tube, adjust it so that its pins fit into the slots and turn the tube by 90°.

Interference[6]

An operating fluorescent bulb can cause considerable radio frequency interference (RFI) to radio and television receivers. The cure is to separate the receivers and the fluorescent bulb as much as possible. It is inadvisable to install fluorescent fixtures above radio or TV sets.

[6](*NEC: Article 410–30b; page 366*)

Efficiency of Fluorescent Lights

Because fluorescent lights generate much less heat than incandescent lamps, they have a much more desirable ratio of watts per lumen. As an example, a 40-watt fluorescent produces about 2800 lumens. To determine the lumens per watt, divide 2800 by 40, and the result is 70 lumens per watt. Consequently, a 40-watt fluorescent will produce about six times as much light per watt as a 40-watt incandescent light.

SL-10 EARTH LIGHT

This miniature fluorescent bulb can be used indoors and outdoors if it is protected against the weather. It is intended for use with table lamps or ceiling fixtures. The SL-10 supplies 75 watts of light, but its efficiency is such that it requires an input of only 18 watts. Its efficiency is superior to ordinary fluorescent bulbs, and its energy consumption is much less than that of a standard 75-watt incandescent type.

Its life span is estimated to be 10,000 hours, an operating life that is more than 13 times greater than a 75-watt incandescent. Its color output is a warm, high-quality light. Unlike fluorescent tubes, it does not need a special fixture and can be screwed into a standard socket of the type used by incandescent bulbs.

HALOGEN LIGHTS

A halogen light (Figure 9–6) consists of a thick glass bulb wall designed to resist breakage. Unlike incandescent bulbs, the halogen light has a flat top, a stand-on-end design. The bulb encloses a capsule that contains a filament made of tungsten. When this filament receives an electric current, it causes the enclosed halogen gas to glow. A halogen gas consists of any one of a group of five chemically related nonmetallic elements that include fluorine, chlorine, bromine, iodine, and astatine.

Because of their structure, these lamps are also known as quartz-halogen bulbs. As in the case of incandescent bulbs, the surface of the filament evaporates, but the halogen gas causes the filament vapor to be redeposited. This extends the life of the filament and also permits the use of a higher temperature.

Unlike the tubular fluorescent, the halogen bulb fits the regular sockets used for incandescent bulbs. Thus, it can be connected as a direct replacement for the more common incandescent.

The halogen bulb has a number of advantages. It will last three times longer than an incandescent, supply 10% more light, watt for watt, and supply excellent light for critical seeing tasks. Using a 60-watt bulb on a comparison basis for both an incandescent and a halogen type, if the average life of an incandescent is 1000

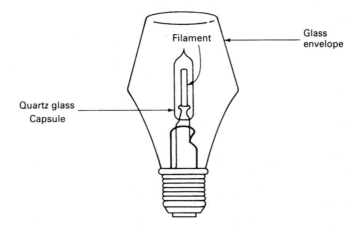

Figure 9–6.   Halogen lamp.

hours, the halogen bulb will last 3000 hours. If the 60-watt incandescent bulb supplies 890 lumens on average, the halogen bulb will supply 980. Another advantage of the halogen is that it gives a whiter, more natural light. It also stays brighter over more of its lifetime. Unlike fluorescent fixtures, it does not need accessory equipment, such as a ballast, nor does it require special supporting fixtures.

Caution: Do not use halogen bulbs in wet locations. Unlike incandescent lamps, halogens may continue to light after the external glass enclosure has been damaged as a result of exposure to moisture droplets or physical abuse. If this occurs, replace the halogen bulb promptly since the inner glass capsule operates at high temperature and pressure and could shatter unexpectedly, creating a risk of property damage or personal injury.

Halogen bulbs have a characteristic not found with incandescent lights. The light they supply can be more readily focused. Because of their greater brightness and focus, they can supply more light in fixtures such as desk types used for sewing, reading, studying, typing, and so on, than incandescent light used for background illumination.

Halogen bulbs are available in many shapes and so can be used in fixtures ordinarily housing incandescents.

HIGH-LINE VOLTAGE VERSUS ELECTRIC LIGHT BULBS

Both incandescent and halogen lights can have a shortened life with high-line voltages. Incandescents and halogen lights can be operated with dimmers, whose effect is to reduce the line voltage received by these lamps. Incandescents are not affected and, if anything, will have extended life. Halogen bulbs depend for their lifetime on filament renewal, and so these lights should be operated on full line voltage for at least 1 hour once a week.

ELECTRIC BILLS VERSUS ILLUMINATION

The electric bill of an average home will include about $70 per year for illumination. This amount is highly variable since the greater the number of bulbs, the higher their total wattage rating. The efficiency of the bulbs and the living habits of the residents are the final determining factors. The greater the efficiency of the light bulbs, the lower is the cost, with a possible reduction of almost 50% in lighting costs.

There are a number of ways of controlling lighting expenditures: keep incandescent bulbs clean; turn lights off when not needed; use larger-wattage bulbs in place of smaller-wattage types, when possible; and use fluorescent bulbs in place of incandescent types.

OUTDOOR LIGHTING[7]

Before starting work on any outdoor lighting installation, check the National Electrical Code. It is also advisable to consult your local code. Soil conditions may be a factor in certain areas, possibly requiring the use of conduit in addition to lead-covered cable. In some instances, plastic cable may be used instead of lead.

One important factor in installing outdoor wiring is having a good ground. For this reason the cable, a UF (underground fused) type, is supplied with a grounding wire. The term, underground fused, sometimes conveys the wrong impression that it is somehow equipped with a fuse. What it means is that there must be a fuse or circuit breaker inside the dwelling at the start of the wiring installation.

Installation

The minimum placement depth for UF cable should be at least $1\frac{1}{2}$ feet, but 2 feet is preferable to avoid accidental spading. Associated equipment such as switches, receptacles, and electric boxes, when mounted partially or entirely outdoors, should be weatherproof types. The surface-mounted receptacle must have hinged covers, with these independently operated for dual-receptacle types. The electric box can be flush mounted on a wall surface facing the outside or on some support such as a post.

[7]*(NEC: Article 410–57; page 371)*

Outdoor Bulbs[8]

Some bulbs can be used either indoors or out. Thus, an outdoor lantern can be equipped with either type. Some bulbs, however, are specifically designed for outdoor use. These include the following:

Low-voltage (12 or 24 volts) incandescent
Mercury vapor
Metal halide
Low-pressure sodium vapor
High-pressure sodium vapor
PAR bulbs

Outdoor bulbs can be controlled in several ways: by manually operated switches, automatically by electric timers, or automatically by photocells.

Since outdoor lights are kept on throughout the night there may be some justifiable concern about the cost. Some light bulbs are specifically designed for long-term use but much more inexpensively than comparable bulbs intended solely for indoors. Such lights are referred to as high-intensity discharge (HID). As an example, a representative HID light will be brighter than a total of five 40-watt indoor bulbs, with all of these bulbs turned on at the same time. The single lamp in this comparison will be a 35-watt HID. Unlike indoor bulbs that are vacuum types, the HID uses a vapor, such as sodium or mercury.

The disadvantage is that the HID is initially more expensive than indoor bulbs, but it not only supplies more light but also has a longer operating life. An incandescent floodlight has a working life of about 1,000 hours, compared to an HID of 20,000 hours.

The light emitted by mercury and sodium vapor types has a color tint, and so one color may be preferred to another. Of the two, the sodium vapor type is more efficient than the mercury. Still, the mercury vapor HID is more efficient than vacuum incandescent floodlights.

Voltage Requirements

Either the line voltage, 120 volts, or 12 volts ac is needed for outdoor lighting. Power-line voltage is needed for post lanterns, floodlights, or automatic-turn-on security lights. Line voltage is supplied directly from a power line inside the home. For low-voltage, a transformer is used to step down the power-line voltage

[8](NEC: Article 410–4a; page 358)

to 12 volts. If both types of lighting will be used, separate lines will be required for the low and higher voltages.

Transformer Requirements

The transformer for step-down voltage is determined by two factors: the amount of load and the distance of the load from the secondary winding of the transformer. To determine the amount of load, add the wattage ratings of each of the lights. These lights are wired in parallel, and so the larger the number of lights the greater the load. The amount of load also determines the wire gauge to use.

Conduit Requirements[9]

There are various types of conduit designed for outdoor wiring. Some of these can be bent with the help of a conduit bender. The advantage of the conduit bender is that the conduit can be made to assume any angle between 0° and 90°. The dimensions of conduit are specified by its length and inside diameter (ID).

Installing Outdoor Lights for Holiday Use[10]

Don't be overconfident. Using outdoor lighting has certain hazards not encountered indoors. When getting ready to string outdoor lights, look up. Don't raise ladders near the electrical service drop coming into a house or business. Electrically nonconductive ladders may be heavier than aluminum types, but they are much safer.

When stringing lights in trees, make sure there are no power lines near the tree. Branches or entire trees can become energized if they contact a power line. Power-line voltage is much higher than the 120/240 volts used inside a home or business. Tree branches may have a moist surface and their internal sap makes them conductive.

If using a pole or other extension to hang outside lights or ornaments, do not use it near a power line. Especially avoid metal poles or poles having a metallized surface. If using a wooden pole, make absolutely sure it is dry. Wear thick, dry gloves.

Never use electric lights on a metallic tree. Use colored spotlights but never have them in contact with the tree.

Check all cords and lights for cracked insulation, frayed or bare wires, or loose connections. Discard damaged cords; don't try to repair them.

[9] (NEC: Article 370–16c; page 305)
[10] (NEC: Article 410–4b; page 358)

Installing an Outdoor Receptacle[11]

Having at least one outdoor receptacle is not only a convenience when using electrical equipment but a necessity. The electrical work for doing so is fairly easy, for all that must be done is to tap into a convenient middle of the run indoor receptacle. The mechanical work, providing a path through a wall for a cable, is more difficult.

To have power available on an outside wall means having a receptacle especially designed for outdoor use. It should also have a snap cover plate to protect the receptacle against the weather. The electric box is a metal type that is weatherproof and equipped with a sealing gasket made of rubber or foam. The object is to provide protection for the contents of the box against the weather. In addition, you may need a supply of threaded metal plugs to cover any unused holes in the electric box.

The wire to use should be UF (underground fused). This wire has an outer covering of heavy plastic, is available in No. 12 or 14 gauge, and is equipped with a black and white wire, plus a ground lead. Alternate wire choices could be type NM or cable specified by your local electrical authority. Still another possibility is type TW wire plus conduit. TW wire is covered with a thin layer of thermoplastic insulation and, although it is water resistant, should be encased in conduit. Some local codes require that outdoor wiring be done by a professional electrician.

The GFCI.[12]   The outdoor receptacle should be equipped with a GFCI as indicated in Figure 9–7. Before wiring the outdoor receptacle, apply caulking compound to its rear frame. Bring the cable through the receptacle and then apply additional caulking compound around its four sides so as to form a seal between the box and the building. This can be done with a caulking gun or applied with a flat-blade screwdriver. Hold the box and drill through the holes so as to install screws for fastening the box to the building wall. At this time, connect the wires from the GFCI and the cable wiring from the indoors receptacle. Connect the wires by wrapping their ends tightly using a pair of gas pliers. Then cover the connections with wire nuts and cover with plastic tape. After mounting the cover plate, restore electrical power by reinserting the branch fuse or resetting the branch circuit breaker. Make sure the spring-loaded cover works properly.

Ideally, the proposed outdoor receptacle and the indoor receptacle from which power will be tapped off should be back to back. Consider also installing another indoor receptacle to house a switch to permit turning off the outdoor re-

[11]*(NEC: Articles 240–32; page 116, 210–52, page 65, 410–57; page 371; 550–8e; page 637; 680–6a; page 757)*

[12]*(NEC: Article 210–8; page 57; 215–9; page 68)*

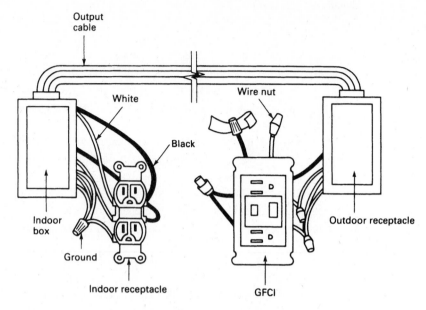

Figure 9–7.   Outdoor receptacle equipped with a GFCI.

ceptacle. This would be helpful if the outdoor cable is to be used to operate a post-top lantern, as in Figure 9–8. This arrangement uses an automatic time switch in place of the on-off switch. This branch line is also used to operate a light fixture and a wall bracket fixture. This circuit does not use an outdoor receptacle, but it will require an outdoor cable for all wiring following the automatic time switch.

Having one or more outdoor receptacles is essential for the electrical operation of units such as a lawn mower, lawn edger, soil tiller, leaf blower, and vacuum cleaner. A great advantage is the elimination of gasoline fuel storage. However, the use of electricity also has its disadvantages. It results in a trailing line cord, which can be a nuisance and possibly a hazard as well. The line cord must have a current-carrying capacity exceeding the rated capacity of the most current-hungry of all the units to be operated.

There are other considerations. The cord must be designed for outdoor use and when not required must be properly stored indoors.

The outdoor receptacle should have a GFCI (ground fault circuit interrupter). As an additional protection, do not use it when the ground is covered with dew or immediately following a rainfall. Wearing dry gloves and waterproof shoes is desirable.

Make sure that the receptacle to be installed is an outdoor type. The source of power can be a connection made to an indoor receptacle, preferably one that is

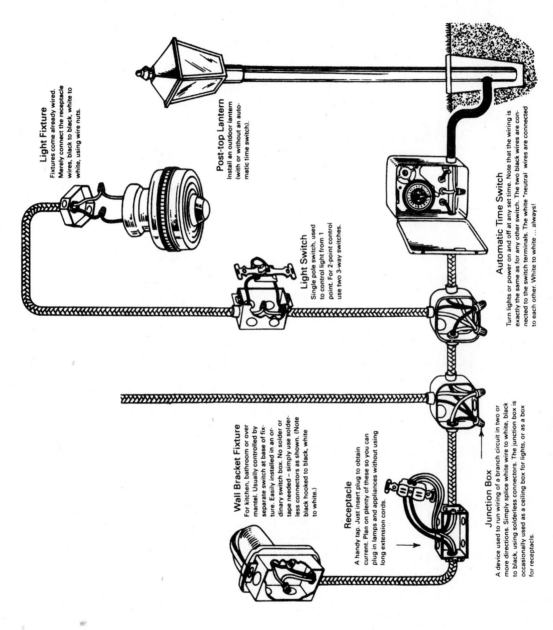

Light Fixture

Fixtures come already wired. Merely connect the receptacle wires, black to black, white to white, using wire nuts.

Post-top Lantern

Install an outdoor lantern (with or without an automatic time switch).

Light Switch

Single pole switch, used to control light from 1 point. For 2-point control use two 3-way switches.

Automatic Time Switch

Turn lights or power on and off at any set time. Note that the wiring is exactly the same as for any other switch. The two black wires are connected to the switch terminals. The white neutral wires are connected to each other. White to white ... always!

Wall Bracket Fixture

For kitchen, bathroom or over mantel. Usually controlled by separate switch at base of fixture. Easily installed in an ordinary switch box. No solder or tape needed – simply use solderless connectors as shown. (Note black hooked to black, white to white.)

Receptacle

A handy tap. Just insert plug to obtain current. Plan on plenty of these so you can plug in lamps and appliances without using long extension cords.

Junction Box

A device used to run wiring of a branch circuit in two or more directions. Simply splice white wire to white, black to black, using solderless connectors. The junction box is occasionally used as a ceiling box for lights, or as a box for receptacls.

Figure 9–8. Wiring arrangement for a post-top lantern

at the end of a branch line, first making certain that the new load will not overload that line.

Voltages for Outdoor Use

The operating voltage for the post-top lantern is the same as the voltage used indoors and is 120 volts. However, a 12-volt system is used for path and shrubbery lighting, with these wired in parallel. The components required for such a system consist of a step-down transformer equipped with a 120-volt primary and a 12-volt secondary. The transformer is usually mounted indoors since the 120 volts required for the primary is readily available. With the transformer positioned indoors, there will be no need for weatherproofing it.

The output power capability of the transformer will be determined by the total number of lights used and their power rating. The total power will be somewhere between 150 and 250 watts. For a 150-watt installation, use 16-gauge cable; for 200 watts, use 14-gauge; and for 250 watts, use 12-gauge.

The wire for connecting the 12-volt lamps can be type UF or TW with conduit frequently specified for TW. UF wire can be used without conduit. For underground wiring it is advisable to check municipal regulations in the event that special soil conditions apply. The wire should be buried to a depth of $1\frac{1}{2}$ to 2 feet, with the greater depth preferable. This is to avoid any cut-through by a tiller.

Types of Outdoor Lights

Well light.   This 12-volt outdoor light is so called since it is designed to be buried in the ground and shine its beam straight up into the air, as indicated in Figure 9–9(a). While it can be used to emphasize tree structure, it can also call attention to house numbers. The well light can be used for dramatic effect during a rain or snow storm.

Mushroom lights.   These lights (parts b and c), can be used to highlight areas of low foliage, borders, walkways, paths, and decorative ground covers. Mushroom lights are available with small and large shades. The 10-inch-diameter shade is ideal for confining the light to a smaller area, while the 14-inch shade illuminates a broader area such as patios or gardens. The mushroom fixture concentrates its light in a downward pattern.

Entrance lights.   The purpose of an entrance light (part d), is to make front doorsteps more visible and to make it easier to use a front-door key. It can also minimize the hazard of objects left outdoors near a walkway or steps. Entrance lights can also feature automatic photo control for dusk-to-dawn operation.

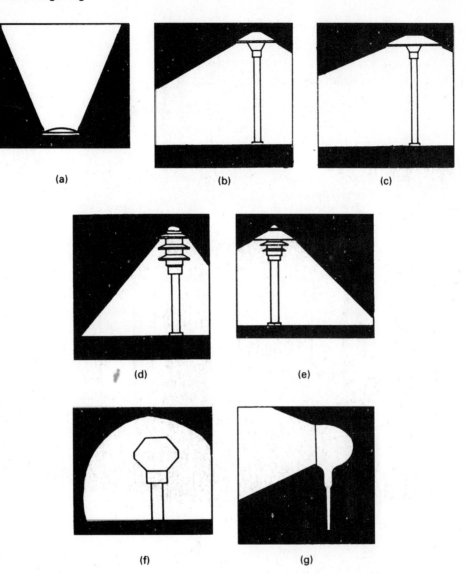

Figure 9–9. (a) Well light; (b) 10-inch mushroom light; (c) 14-inch mushroom light; (d) Entrance light; (e) Tier light; (f) Globe light; (g) Floodlight.

Tier lights.   These are available in both plastic and metal fixtures. There are two advantages of the plastic type. It will not corrode and its color is built into the fixture and will remain permanently. The tier arrangement (part e) casts its light downward in a soft ring of illumination.

Globe lights.   The purpose of a globe light is to use an outdoor 12-volt type to supply light coverage for a large area. It is commonly used around the perimeter of swimming pools, hot tubs, porches, and recreational areas (part f). Two styles of globe lights are available. One has a translucent frosted globe, while the other features a shaded globe. The latter type provides a more downward directed lighting effect.

Floodlights and spots.   A floodlight (part g), sometimes referred to as a flood, and a spotlight, often called a spot, are outdoor lights used for supplying a light having a high intensity and covering a large area. The difference between these two is that the spot does not supply as much light and requires less operating power. It emphasizes light on a smaller area.

Floods and spots can consist of individual bulbs, but are also available in a cluster of two or three mounted on a swivel so as to be able to direct the light over a certain area. A reflector is generally used so as to concentrate the light in a forward direction.

Floods and spots are current-hungry devices and so become extremely hot very quickly. Give the bulbs ample time to cool before trying to remove them after use. To determine the amount of current required by these incandescent types, divide their power rating in watts, marked on the bulb, by their applied voltage, an emf of 120 V.

If an extension cord is to be used with floods or spots, make sure it is an outdoor type and that it has a current rating equal to that of the flood or spot, or preferably higher.

Many types of low-voltage floodlights are available. Some use a Par 36 sealed beam automotive lamp. Another popular style features a round lamp in a metal or plastic housing. The sonic-sealed lamp used with this floodlight is designed especially for landscape illumination. The lamp is recessed in the housing to cloak the light source.

Still another uses a high-intensity halogen lamp that is whiter and brighter than a conventional floodlight. It provides more light per watt of power. One newer floodlight design features a rectangular lamp that directs the flow of light in a geometric pattern.

Sensor lights.   This is another type of outdoor light, but it works from a 120-volt source. It is equipped with one or a pair of light bulbs designed for outdoor use (Figure 9–10). It is equipped with a motion sensor and is used in the

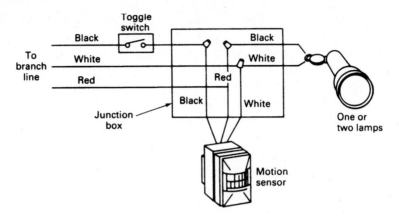

Figure 9–10. Motion sensor light

dark to respond when a person, animal, or large object moves into its vision level. From dusk to dawn it can turn on its electric lights when its photocell is triggered, and it will turn these lights on automatically. When the motion stops, the lights will stay on for a predetermined amount of time and then will automatically turn off. The unit has three lengths of on time: 1, 2, or 3 minutes. No special outdoor-type receptacle is required, for it is supplied with the unit. A gasket is used around the outside edge of the box to protect it from the weather.

The electric box of the sensor fits snugly against a wall. The wiring is led through the wall and connects to a convenient indoor receptacle. No switching is required since the sensor lights turn on automatically.

Scenic Outdoor Lighting

Outdoor lighting helps ensure safety and security but it can also be used to enhance the appearance of the grounds around a home. There are various techniques that are available.

Uplighting. This is most commonly used to focus interest on large trees, statuary, or plants (or to illuminate a leafy canopy). The light sources can be ground mounted or recessed into the ground and directed upward and away from the viewer to prevent glare.

Plants and shrubbery with distinctive shapes or foliage can be emphasized using these uplight variations:

Silhouetting can be achieved by positioning the light fixture behind and below the object of attention, and aiming the light at a wall or fence several feet behind that object.

Shadowing results when the light source is placed in front of a plant and directed through it so that its shadow (usually enlarged) is projected against a rear wall or fence.

Recessed uplights are ideal for flush-mounting in open lawn areas or concrete surfaces. For more flexibility or in cases where seasonal changes may dictate re-aiming of light sources, adjustable ground-mounted fixtures would be more suitable.

Downlighting.   In contrast to uplighting, downlighting exhibits a more natural and subdued quality. Actual effects will vary depending on the height of the light source, and angle of direction. Stem-mounted fixtures can be used close to the ground to highlight specimen plants or provide directed illumination of walkways or paths. Higher mounting of sources (e.g., under the eaves of a house) provides general lighting for security. In addition, 'moonlighting' effects can be had by mounting the light source high in the trees. The result is a soft, natural filtering of light, which illuminates foliage and lawn areas, while creating shadow patterns on ground surfaces.

Accent lighting.   Signs, statuary, specimen plants, or other interesting features can be isolated and emphasized through the judicious use of high-intensity spot lighting. Ideal choices here are shielded, tree-mounted light sources or recessed ground features, which can be hidden from direct view by shrubbery or ground plants.

Wall washing/relief lighting.   Walls and building exteriors are natural candidates for inventive lighting techniques. Relief lighting adds depth to any textured surface, while smooth walls can be sculpted using directed sprays of light. Grazing is a form of relief lighting in which the light fixture is placed several inches away from a textured wall or facade and angled slightly to better define boundaries, contours, or irregularities.

Scalloping.   Different lighting effects can be had by mounting the light source farther back from the surface, up to several feet. At source distances of 10 to 15 feet, the net effect is a general illumination for flooding, ideally suited for defining boundaries and adding perspective to the scene.

Spread lights are designed to produce circular patterns of illumination for visibility, security, and accent lighting. Size and intensity of the spread area is determined by the height of the source fixture.

Chapter 10

DC Motors[1]

Motors are sometimes compared to transformers since both have an electrical input. The transformer has a fixed-position secondary winding, but the comparable winding in a motor is designed to rotate. An important difference is that the motor is a transducer and changes the electrical energy it receives to mechanical energy. In the case of the transformer, the input is electrical energy and so is the output.

IDENTIFYING MOTORS

Motors are often identified by the work they do and so may be referred to as sewing machine motors, vacuum cleaner motors, clock motors, and so on. However, this is limited knowledge for it tells us very little about the motor.

Motors are often categorized in two ways: by the kind of energy input, whether dc, ac, or ac/dc, and by some other electrical or mechanical characteristic. DC motors can be described as:

Shunt	Differential
Series	Interpole
Compound	Universal
Split field	Tapped field universal

[1](NEC®: Article 430; page 414)

There are a number of variations within this listing. Identifying a motor as a shunt-wound type, for example, simply narrows a choice to one category, for there are a number of different kinds of shunt motors. A further classification would provide a more specific analysis, such as:

1. *Physical size of the motor.* This could include a description such as flea, fractional, or whole number and is expressed in terms of horsepower (hp). Thus, a motor could be described as $\frac{1}{20}$ hp, $\frac{1}{4}$ hp, or 5 hp. As a general rule, the larger the horsepower output of a motor, the greater its size. In the case of ac motors, some manufacturers specify peak horsepower. This is deceptive since peak horsepower occurs only twice during each ac cycle.

2. *Description of its function.* Motors are made to do specific jobs, such as a washing machine motor, a vacuum cleaner motor, and so on.

3. *Electrical power input.* A motor could be rated as 300 watts, 500 watts, and so on. Usually, the greater the power input is, the larger the physical size of the motor.

4. *Mechanical features.* Such as closed frame, open frame, self-cooling, splashproof, explosion proof.

5. *Speed.* Usually indicated in revolutions of the shaft per minute (rpm).

6. *Type of armature rotation.* The speed could be constant, varying, or adjustable. There could also be one or more speed settings.

7. *Type of drive.* The shaft of the motor could be used to operate a set of gears or a belt. It could also be a direct-drive type or operate some kind of flexible shaft.

8. *Motor mounting.* Some motors are fixed-position types, such as washing machine motors. Others are designed to be moved together with the appliance they operate, such as vacuum cleaner motors.

THE NATIONAL ELECTRICAL CODE AND MOTORS[2]

The Code has a number of articles with reference to motor location, motor control circuits, methods of disconnecting motors, grounding fixed-position motors, the use of junction boxes for motors, and so on. The Code uses the 400 series articles for motors, such as 430-14, 430-16, 440, and others.

[2]*(NEC: Article 210–22a; page 61)*

MOTORS AND THEIR PARTS

The concept behind the construction and use of motors is very simple and depends on the forces of attraction and repulsion between magnets. These magnets are of two types, permanent and electromagnet. A permanent magnet consists of a metal or an alloy that is capable of being magnetized. An electromagnet consists of a coil of wire carrying an electric current, either dc or ac. The advantage of the permanent magnet is that, once the magnetizing process is completed, no further electrical energy input is required. Also, modern magnetizing methods and the development of special alloys have resulted in magnets having a high degree of magnetization and a magnet that requires relatively little space. The advantage of an electromagnet is that it can be made to have a desired magnetic strength, with a magnetic field that can be turned on or off.

There are two basic parts to a motor, an armature, or rotating element, and a stator, a part that is fixed in position. Both the armature and stator are magnets and can be permanent or electromagnets or some combination of the two. The armature is mounted on a shaft supported by bearings that permit the armature to have easy rotating ability. It is the armature that does the work of the motor.

The stator, also called a field coil, receives its operating voltage directly, but the armature presents a problem and that is how to supply it with an electric current since it is in motion.

BASIC DC MOTOR[3]

Figure 10–1 shows the setup of the basic dc motor. As indicated in drawing a, the motor consists of two magnets, one of which is a permanent type and the other an electromagnet. The permanent magnet has a pair of fixed poles, a north (N) and a south (S). A steady magnetic field exists between these poles.

The rotor, an electromagnet, is positioned in the space between the poles of the permanent magnet. The rotor coil, consisting of a large number of turns, is mounted on a shaft, and so the coil is free to rotate. The ends of the coil are permanently connected to a pair of metal rings that are insulated from each other, which is referred to as a *commutator*. The commutator is positioned on the rotor shaft and turns with it. Riding on the commutator are a pair of electrically conductive brushes, usually made of carbon. The brushes, which are fixed in position, make good electrical contact with the commutator. The commutator is connected to a dc voltage source, represented in the drawing by the symbol for a

[3](*NEC: Table 430–147; page 453*)

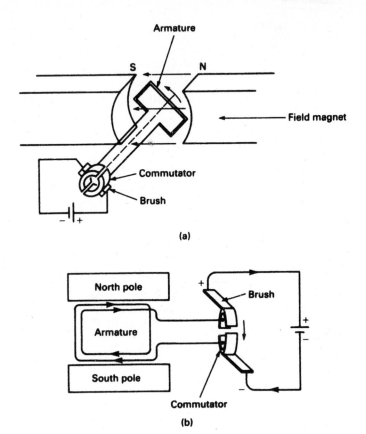

Figure 10–1. Two views of the basic dc motor.

battery. The direct current flows from the negative terminal of the voltage source to its nearest brush, from the brush to the commutator, from the commutator through the armature coil, and then to the other half of the commutator. From here the current passes through the brush to the positive terminal of the dc voltage source. The current continues through that source to the negative terminal, and from here the entire process is repeated. Drawing b supplies a different view. However, its operating features are the same.

The arrows show the direction of flow of the current, and since the current is dc, the current flow is always in the same direction. As this current passes through the armature coil, it makes it into an electromagnet. The magnetic field of this electromagnet reacts with the magnetic field of the permanent magnet. The magnetic force is one of repulsion, forcing the armature coil to turn.

The amount of turning force depends on a number of factors: the number of turns of wire forming the armature coil, the amount of current flowing through it, the distance between the poles of the permanent magnet, the strength of the permanent magnet, the amount of friction of the armature shaft, the electrical resistance of the brushes, and the kind of contact between the brushes and the commutator.

CLASSIFICATION OF MOTORS BY OPERATING SPEED[4]

To get a more complete classification of a motor type requires a somewhat detailed listing of both electrical and physical characteristics. One of these is the motor's operating speed.

1. *Constant speed.* A constant-speed motor is one that maintains its rpm despite changes in its operating load. This does not apply to a condition in which the load is completely removed. Some motors will race in the absence of a load; others will not be affected. Thus, constant speed is a generalization and is affected by the design of the motor.

2. *Controlled angular rotation of the armature.* Some motors, known as steppers, can have a predetermined angular rotation of the shaft. The rotation may be partial, such as 20°, 30°, or more, or may be a complete but single revolution. The stepping action is automatic and can be made repetitive. Once set, the amount of rotation is not under the control of the motor operator.

3. *Varying speed.* Some motors are sensitive to the amount of loading, and as this increases the rotational speed decreases. With a decreasing load, the armature may have a tendency to race.

4. *Adjustable speed.* Motor speed can be used with this motor to increase or decrease the motor armature's rpm. Once a desired operating speed has been reached, increasing or decreasing the load will have no effect.

5. *Adjustable varying speed.* This motor has a speed operating characteristic that is similar to the adjustable-speed type. However, in this case the motor speed can vary if the amount of load is changed.

6. *Multispeed characteristic.* This motor has a number of operating speeds set by a separate control. Some motors, such as ceiling fan types, are positioned so that a speed adjustment is not readily accessible if mounted on or near the motor. A motor speed control can be placed in a standard wall-type electric box for multispeed operation. The control box does not indicate the actual rpm of the motor, but does so in a general sense, such as low, medium, and high.

[4]*(NEC: Article 430–7a; page 417)*

PARTS OF A DC MOTOR

The part of a DC motor are as follows:

1. Frame
2. Armature
3. Laminated core on which the armature is wound
4. Shaft
5. Field magnets, either permanent magnets or electromagnets
6. Commutator
7. Brushes

Armature[5]

The armature consists of a number of coils of wire wound in rectangular form, with these coils embedded in the slots of a laminated, circular core. The core, which is made of a ferrous material, not only supports the coils but adds to their inductance. As an electric current flows through these coils, they become electromagnets and are surrounded by a strong magnetic field.

As indicated in Figure 10–2, the ends of the coils are connected to metallic bars that form the commutator. Each coil is connected to a pair of commutator segments. Figure 10–3 shows how the armature coils are connected to the commutator bars.

The complete assembly of the armature and the commutator is mounted on the shaft. The shaft supports this assembly and by a ball-bearing arrangement permits the armature and its associated commutator to rotate (Figure 10–4).

When an electric current flows through the armature coils, it becomes an electromagnet. The problem of supplying a current to a moving coil assembly is

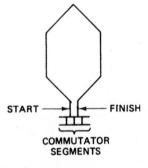

Figure 10–2. Although shown here as a single turn, each armature coil consists of a number of turns. The ends of the coils are connected to a commutator segments.

[5](*NEC: Article 430; page 414*)

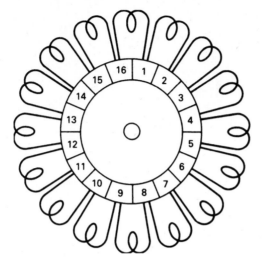

Figure 10–3. Structure of the armature and commutator.

solved as indicated in Figure 10–5. An electric current supplied by a dc source is connected to a brush tension spring. It then continues to a brush mounted in a brush holder. The current flows through the brush and then into some commutator segments. The current moves only to those commutator segments with which the brush makes contact. But since the commutator is mounted on a rotating shaft, the coils all receive current in turn. As each coil receives current it becomes an electromagnet and so is surrounded by a magnetic field.

Brushes

Brushes are held in devices known as brush holders. The face of the brush rests on the rotating commutator, with an electric current flowing through the brush into commutator bars. Brushes are made of carbon treated to have different electrically conductive properties. These brushes are identified as hard carbon, electrographitic carbon, graphite, and metal-graphite.

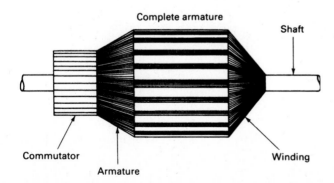

Figure 10–4. Armature and commutator assembly.

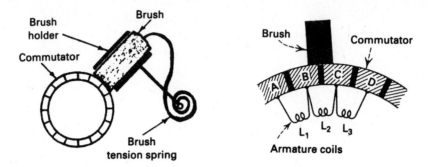

Figure 10–5. Two views of how the commutator receives current via the carbon brush.

Commutator

The commutator consists of rectangular lengths of hard-drawn or forged copper that are insulated from each other, form a circle around the shaft, and consequently rotate with it. Each armature coil makes connection with a pair of bars.

Field Coils or Permanent Magnets

A motor works because of the reaction between the magnetic fields supplied by the armature and by the field coil or permanent magnets. Unlike the armature, the field coil is fixed in position and so it can be connected directly to a dc voltage source. The armature rotates in the space between the poles of the field coil arrangement. Figure 10–6 shows the positioning of the field coils on a laminated ferrous core.

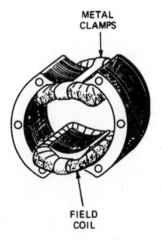

Figure 10–6. The armature fits in the open space between the field coils.

CLASSIFICATION OF MOTORS BY THEIR FRAMES[6]

The interior of the frame of a motor is used to support the various components of which the motor is made, notably the stator and the rotor. But another function of the frame is to protect the motor from dust, water, and the possibility of an explosion. The motor frame is also known as an *enclosure* or *yoke*. It not only acts to protect the motor parts but also works as a heat sink, transferring heat developed inside the frame to the outside and thus preventing excessive heat buildup. The base of the motor is used as a support.

The frame of the motor also has electrical characteristics. It works somewhat in the nature of a shield to minimize radiation of electrical interference. Because of its ferrous nature, it adds to the inductance of the field coils.

Frame Types

Open frame.[7] The open frame has large openings to permit the escape of heat to the exterior. In some instances the motor is also air cooled through the use of a blower.

Totally enclosed frame. In some instances the frame encloses the motor completely (Figure 10–7). This prevents poking fingers inside the motor when the rotor is still functioning. It makes the motor more environmentally safe by protecting it against smoke, dust, and chemical fumes.

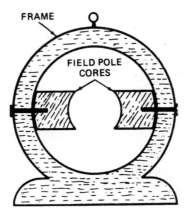

Figure 10–7. Totally enclosed motor frame.

---

[6](*NEC: Article 430; page 414*)
[7](*NEC: Article 430; page 414*)

Protected frame. This is a modification of the totally enclosed frame. It has vents, but these are protected by screening. This helps protect against accidentally touching rotating parts, but offers less protection against fumes and dust. It can be used in environments that are not hostile. The vents may be on the front of the motor near the exit portion for the motor shaft. If the motor is internally cooled by a fan, it may have vents around the edges of the frame.

Drip-proof frame. While this frame protects the motor against liquid dropping it has no provision for ventilation. The assumption is made that drips falling on the motor at any angle not greater than 15° off the vertical cannot enter.

Splash-proof frame. This motor enclosure is designed so that liquid or solid particles falling vertically or reaching the frame at some angle not greater than 100° from the vertical cannot get inside the motor.

Dust-proof frame. This frame encloses the motor completely to prevent the entry of dust from any angle. It is similar to the totally enclosed frame.

Watertight frame. This frame is so designed that it encloses the motor completely, so much so that a spray of water applied from a hose at any angle cannot find its way into the interior of the motor.

Explosion-proof frame. This frame is similar to the totally enclosed type and can withstand an internal motor explosion. A motor explosion is not to be attributed to the motor but rather to the presence of some gas or dust in the immediate environment of the motor. This motor, as well as those previously described, may have a frame that is ribbed externally to function as a heat sink to provide a greater heat-radiating surface.

SERIES-WOUND DC MOTOR

This motor is called series wound since its field and armature windings are in series with each other. As indicated in Figure 10–8, it uses electromagnets for both the field and armature windings. It is called dc since its voltage supply is from a dc source. This voltage supply can be supplied by batteries, a dc generator, or an electronic power supply having a dc output. The type of supply used is determined by the current requirements of the motor. Whichever supply is used, it is referred to as the line voltage.

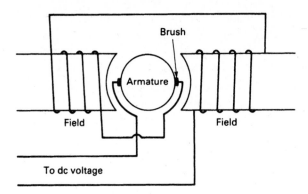

Figure 10–8. Series-wound motor.

Characteristics

The series-wound dc motor is used primarily for fluctuating loads, and even if the motor has a heavy overload, it can be started. However, the motor has poor speed regulation, indicating that its speed is affected by the amount of overload. Thus, the greater the load the slower the speed. The speed can be increased by lightening the load. Variable resistor R shown in series with the line in Figure 10–9 is used as a speed control. Since this resistor is in series with the armature and the field windings, it must be capable of carrying the current required by these components. The motor is suitable for repeated acceleration with heavy loads. In the event the load is removed, the motor will race, assuming the line voltage remains connected.

The motor torque, or its turning force, varies approximately as the square of the field or armature current. The torque must not go below 10% to 15% of the full-load torque.

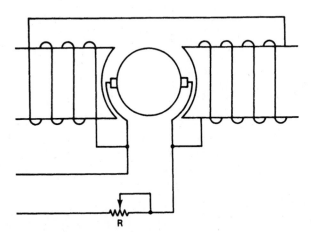

Figure 10–9. Series-wound motor with speed control, R.

Most dc motors are equipped with interpoles. An *interpole* is a supplementary magnet sometimes referred to as a commutating pole. The purpose of an interpole is to minimize the sparking that occurs between carbon brushes and the commutator bars. The interpoles are wired in series with the field and armature coils.

The direction of rotation of the armature can be changed by transposing the leads to the dc voltage source. This reverses the direction of current flow.

The field winding is constructed of a few turns of heavy wire, that is, wire having a fairly low gauge number. The strength of the magnetic field around the field coils depends on the line current and the iron core around which the field is wound. This motor is characterized by having excellent starting and stalling torque.

SPLIT-FIELD SERIES-WOUND MOTOR

Often, what appears to be a completely new motor type is simply an existing motor with some modification. The modification does not necessarily mean a completely new design, but can be just a simple change. The split-field dc motor falls within this category. It is referred to as split field because, as shown in Figure 10–10, it has a tap at the electrical center of the field coil. This tap, plus a single-pole, double-throw (SPDT) switch is an easy way to get a change in the direction of rotation of the armature. This is easier and faster than changing the connections to the dc power source. All the characteristics of the series-wound dc motor are applicable.

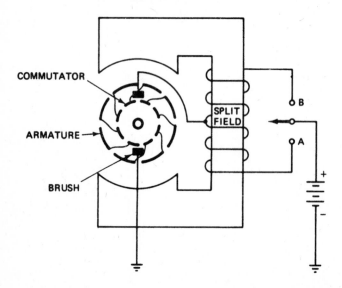

Figure 10–10. Split-field series-wound dc motor.

SHUNT-WOUND DC MOTOR

The connections of the shunt-wound dc motor are the opposite of those of the series-wound motor. The armature coils and the field coils are connected in parallel, with both wired directly across the dc power source, as in Figures 10–11a and b. The field coil is wound with many turns of fine wire, and the amount of current flowing through it is not the same as the armature current, but is actually much less. However, it has a strong magnetic field due to its large number of *ampereturns,* that is, amperes multiplied by the number of turns.

Characteristics

The total line current taken by this motor from its dc source is equal to the sum of the field and armature currents. Even with a varying load, the rotational speed of the armature remains fairly constant.

A characteristic of all motors is that they develop a voltage in opposition to the source voltage. Known as a counterelectromotive force or counter emf, at full armature rotational speed, this counter emf is almost equal to the source voltage. The production of a counter emf means that the motor also works as a generator.

The speed of this motor can be controlled (*NEC: 430-98a and b; page 446*) in several ways. One way is by putting a variable resistor in series with the field coil, thereby adjusting the amount of current passing through it. Another is to adjust the source voltage. Doing this varies the strength of the magnetic field around the field and armature coils. An increasing current from the source makes these magnetic fields stronger; a decreasing current makes them weaker.

The starting torque of this motor is less than that of the series wound. For

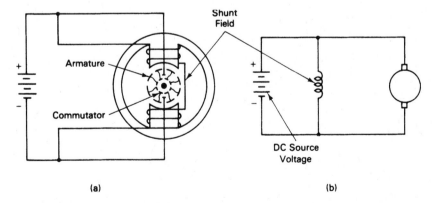

Figure 10–11. (a) Shunt-wound motor and (b) its circuit diagram.

this reason the series wound is preferred when a heavy load is to be started from a rest condition. When the shunt motor is to be started, the field winding should receive its excitation before that of the armature. If, for some reason, there is an open in the shunt field, the armature speed will rise substantially.

The armature current and resistance and the field current and resistance are related to each other. If any three of these quantities are known, the fourth unknown value can be determined from these formulas.

$$I_1 = \frac{I_2 \times R_2}{R_1}$$

$$I_2 = \frac{I_1 \times R_1}{R_2}$$

$$R_1 = \frac{I_2 \times R_2}{I_1}$$

$$R_2 = \frac{I_1 \times R_1}{I_2}$$

where

I_1 = armature current
R_1 = armature resistance
I_2 = field current
R_2 = field resistance

Because of the large number of turns of the field winding, it has a substantial inductance. It is true that the input is dc but the current is a growing one and so the field exhibits a certain amount of reactance. This reactance disappears when the current reaches and sustains its maximum amount. It takes time to get to this level, thus it is desirable to prevent application of power to the armature until the field current is maximum.

COMPOUND-WOUND DC MOTOR

The compound-wound motor is a compromise between the series- and shunt-wound motors and has some of the characteristics of each. Like the series wound, it has a large starting torque. Its operating speed remains fairly constant not only for a fixed load but also when quick changes are made between loads, such as going from a light to a heavy load. However, the no-load speed is very low. If the load is completely removed or becomes very light, there is no danger of armature speed runaway.

This motor is equipped with both series and shunt field windings. Figure 10–12a shows the two field windings of this motor. The motor speed is controlled by a wirewound rheostat. Drawing b is the schematic diagram.

The series-wound motor is well suited for operating heavy starting loads such as subway trains or elevators. However, the shunt and compound motors are more widely used, since they have a greater number of possible applications. The compound wound is preferable for larger motor sizes.

Reverse rotation of the compound wound can be had by interchanging the leads at the brush holders. This does not necessarily mean making the change physically, since the motor may be equipped with a reversing switch.

DIFFERENTIALLY-WOUND COMPOUND MOTOR

The differentially-wound compound motor is a member of the compound-wound family of motors. A characteristic of compound-wound motors is that the magnetic field around the shunt winding aids the magnetic field around the series winding. However, the differentially-wound compound motor works in an opposite manner; that is, the two magnetic fields oppose each other. The advantage of

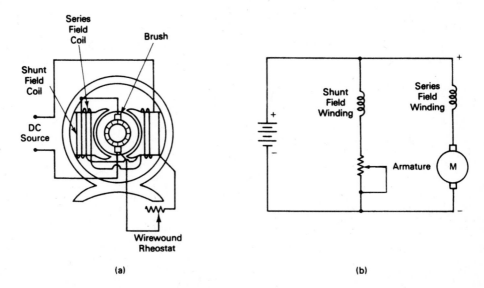

Figure 10–12. (a) Compound-wound motor and (b) its circuit diagram.

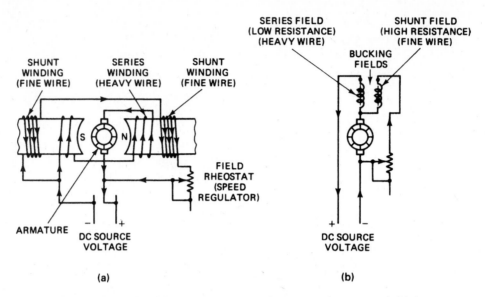

Figure 10–13.   (a) Differentially-wound compound motor and (b) its circuit diagram.

this arrangement is that it supplies automatic regulation of the armature speed under varying load conditions. The speed regulation of a motor, expressed in percent, is the amount by which the rotating armature will vary its speed under varying load conditions. The ideal setup is one in which the motor has 100% regulation. This means that the armature speed will remain constant. Figure 10–13 shows the wiring diagram for this motor.

CUMULATIVE-WOUND COMPOUND MOTOR

The cumulative-wound compound motor is yet another compound-wound type of motor and is illustrated in Figure 10–14. This motor has some similarity to the series wound since it has a high starting torque. However, it does not have the armature runaway characteristic of the series type. In the differentially-wound motor, the currents through the series and shunt fields move in opposite directions; consequently, their magnetic fields are in opposition. In the cumulative-wound motor, the flow of current through the series and shunt fields is in the same direction. Consequently, the magnetic fields aid each other. Series aiding and series opposing magnetic fields are based directly on the behavior of inductors connected in this way.

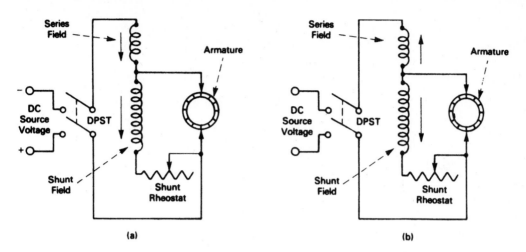

Figure 10–14. (a) Cumulative-wound compound motor; (b) differentially-wound compound motor.

STEPLESS MOTOR

All the motors described so far come under the general heading of stepless; that is, the armature goes through a number of complete rotations, with each rotation being 360°. If the load is very light, such as a small battery-operated fan, the blades of the fan can continue turning for a short time after the power is turned off. The number of turns of the armature is of no consequence, even if the armature can be made bidirectional, in which case it can be referred to as a bidirectional stepless.

STEPPING MOTOR

The rotational characteristic of a stepping motor is opposite that of the stepless, for its shaft can be made to rotate a precise number of turns or can have armature rotation through a selected number of degrees less than 360. Also, the armature can be made to turn through the same number of degrees repetitively over a chosen period of time.

There are various types of stepping motors:

Disk stepper
Variable reluctance stepper
Permanent magnet stepper
Hybrid stepper

The *step angle* is the rotational distance, in degrees, through which the armature turns. The larger the number of steps the smaller the step angle. The step angle can be determined from

$$\text{step angle} = \frac{360}{\text{number of steps}}$$

The number 360 is obtained from the fact that a single complete rotation is always 360° and is applicable to all motors, regardless of size or speed of rotation.

The step angle is referred to as the *step size* and is always specified in degrees. The step angle, for example, can be any value of degrees, consisting of a whole number only or a whole number plus a fraction, provided it is divisible evenly into 360°. Typical stepping angles could be 7.5, 9, 11.25, 153, and 180°. Thus, 360/7.5 = 48 steps, 360/9 = 40 steps, and 360/11.25 = 32 steps.

Step Frequency

The *step frequency* is the number of steps through a single complete revolution of the motor shaft. The step frequency is commonly 24, 32, or 40 steps per second. The second is the basic unit of time used in measuring step frequency. A single step can be further subdivided into smaller angles known as microsteps.

Steppers are not controlled manually, but rather through an electronic drive unit that supplies a drive signal pulse operating at a previously selected frequency.

Unlike stepless motors, which have a continuous, unbroken series of shaft revolutions, the shaft of the stepper rotates in specific amounts or increments. If a single pulse of the drive signal produces a 10° turning movement of the shaft, it will complete 36 increments for one complete revolution.

DISK STEPPER

Figure 10–15 is a drawing of the disk stepper motor. It is equipped with a pair of electromagnets, each connected to a dc voltage source. These electromagnets have a small rectangular opening on one side with a samarium-cobalt disk rotor positioned in the space. The two electromagnets are referred to as the stator poles, with the lower one referred to as phase A and the upper as phase B; hence the motor is sometimes referred to as a two-phase stepper. Each phase is electrically shifted by 90°.

The rotor disk has a number of permanent magnets, with 50 commonly used, although some steppers have more and others have fewer. The number of increments is determined by the number of magnets, and the greater the number of magnets, the smaller the step angle that can be obtained. Each magnet can measure about 1.5 × 1.5 millimeters. The diameter of the disk ranges from 30 to 100 millimeters.

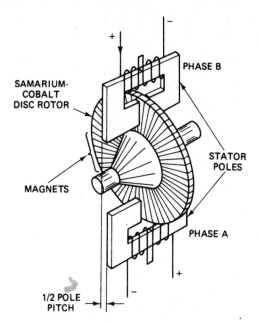

SAMARIUM-
COBALT
DISC ROTOR

PHASE B

STATOR
POLES

MAGNETS

PHASE A

1/2 POLE
PITCH

Figure 10–15. Disk stepper motor. (*Courtesy Penton/PC, Inc.*)

VARIABLE RELUCTANCE MOTOR

Known as a VR stepper, the variable reluctance motor, shown in Figure 10–16, has a toothed rotor mounted on a shaft. Surrounding the rotor is a stator that is also toothed, with each tooth made of a permanent magnet. Two of the stator teeth are electromagnets, with their polarity determined by the direction of current flow through their surrounding coils. The number of teeth on the rotor and stator as well as the number of phases determines the step angle. A step occurs when one phase is de-energized, while at the same time the next phase in sequence is energized. The rotor will then move to a new position of minimum reluctance, with the rotor and stator teeth lined up. This movement completes one step in the motion of the rotor.

ENERGY LOSSES IN DC MOTORS

The input to a dc motor is dc power. In terms of a formula, the power is

$$P = E \times I$$

the power, P, is in watts, E, is in volts, and I is in amperes. The dc motor is an energy converter changing that power at the motor's input to horsepower (hp). There is a relationship between horsepower and electrical power. It will be noted

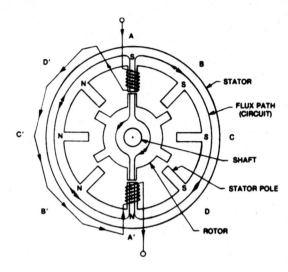

Figure 10–16. Variable-reluctance stepping motor. (*Courtesy Small Motor Manufacturers Association*)

that the electrical power output that is equivalent to horsepower is always less than the electrical power input. This power loss is the sum of a number of individual losses in the motor, some of which are minor; others are major.

Copper Losses

Whenever a current flows through a conductor, such as the armature and field windings of a motor, there are losses based on the amount of current and the resistance of those windings. These losses can be calculated from

$$P = I^2 R$$

Here the power is in watts, I is the current in amperes, and R is the resistance in ohms.

While the resistance of either the field coils or the armature can be very low, the amount of current can be substantial. Also, the loss is proportional to the square of the current.

Power losses are dissipated in the form of heat. For some large-size motors, the heat may be substantial enough to require internal or external cooling. For internal cooling, fan blades are attached directly to the shaft; for external cooling, a separate blower fan is used.

Iron Losses

Iron losses can be divided into two types: hysteresis and eddy-current losses. Hysteresis losses are due to the fact that the armature's iron core revolves in a changing magnetic field. As a result, it becomes magnetized, but with a changing magnetic polarity. The iron or steel of which the armature's core is made always has some residual magnetism, and it is the need to constantly reverse this magnetism that results in energy losses.

Eddy-current Losses

Eddy-current losses are minimized through the use of laminations instead of solid iron cores. Eddy currents heat the cores, and in turn these can cause damage to the windings if their insulation cannot tolerate the temperature increase.

Eddy currents, like currents flowing through windings, result in an energy loss. The thinner the laminations the lower this loss will be. The laminations are also insulated from each other by lacquer or enamel.

Losses due to Friction

Even though the shaft has both its ends riding on a film of oil, there is still a certain amount of friction. This friction is indicated by an increase in temperature. There is some friction between the brushes and the commutator bars on which they ride. There is also some power loss due to current flow through the brushes.

Load Loss

The heavier the load on a motor is, the greater the amount of input power required. Some energy is used in getting the load started, an amount that may decrease or increase depending on the type of load that is used. If the motor is geared, there is some energy loss due to friction in the gear train.

Chapter 11

AC Motors and Generators[1]

The main difference between dc and ac motors is not so much in the motors but in the supply source. In terms of time, dc motors preceded ac types, but simply because dc as a voltage source was well established before ac. DC motors can be operated from dc generators, batteries, or electronic power supplies using ac as an input but having dc as its output.

There are a number of different types of ac motors, but in general they all follow a simple basic design, with the differences due to design modifications. A partial listing of AC motors would include:

Series motors	Universal
Capacitor start	Hybrid
Split capacitor	Single phase
Two-value capacitor	Two phase
Squirrel cage	Three phase
Shaded pole	Repulsion start
Autotransformer	Induction

[1](NEC®: Article 430; page 414)

These are just a few of the many names that are used and, to add to the confusion, the same motor may have two or more different designations.

BASIC AC MOTORS

Figure 11–1 is a drawing of the basic ac motor, and there are some points of similarity to the basic dc motor. This motor has a permanent magnet whose north (N) and south (S) poles are separated to permit the use of a rotor coil mounted on a shaft. Unlike the commutator used on dc motors, what we have here is a pair of metallic slip rings insulated from each other but mounted on the same shaft. Riding on the surface of the slip rings are carbon brushes, which are wired to an ac source. The brushes are held in position by a brush holder, with each holder wired to one end of the rotor coil.

As in the case of the dc motor, the ac motor works through the repulsion of two magnetic fields, one around the rotor coil and the other the magnetic field supplied by the permanent magnet. The polarity of the rotor's magnetic field keeps changing in step with the frequency of the input ac.

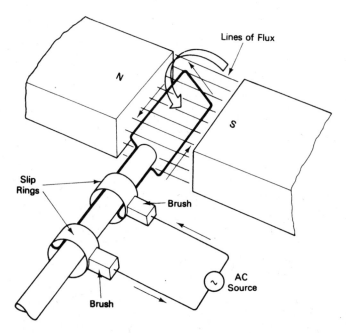

Figure 11–1. Basic ac motor.

BASIC AC GENERATORS

In a basic ac generator (also referred to as an alternator) the revolving coils, known as an armature, turn in a fixed magnetic field supplied by the coils of a stator. This magnetic field can be supplied by a fixed permanent magnet or an electromagnet.

Whether the unit works as a motor or an ac generator depends on the input. As a motor, the input is electrical. As an ac generator, the input is mechanical. The input can be a motor, a turbine, waterpower, windpower, a gasoline engine, or a steam engine, depending on the size and weight of the armature and on the amount of voltage and current wanted from the generator.

When used as a motor the shaft of the unit is equipped with a pair of slip rings. These are connected to the ends of the armature coils and so the slip rings rotate with the armature. The brushes, positioned on the slip rings, make good electrical contact with them, permitting a rather easy connection to an external voltage source.

A generator is an electromechanical device for producing a voltage and a current. It works on the principle of moving a length of wire through a magnetic field. The magnetic field can be supplied by either a permanent magnet or an electromagnet. The length of wire, usually copper, in most cases is wound in the form of a coil.

A voltage is generated (hence the name of the completed device) when the conductor is moved through the magnetic field or the magnetic field is moved across the conductor. When revolving in the magnetic field the coil of wire will do so at some angle ranging from 0 degrees to 90 degrees. At 0 degrees the generated voltage is zero; at 90 degrees it is maximum.

In moving thorough the magnetic lines the action is simply one of movement. The lines aren't cut or bent out of shape. In the process the resulting voltage is said to be induced.

The amount of induced voltage will depend on several factors: the number of turns of the conductor in the form of a coil, the type of iron used in making the laminations on which the coil is wound, the strength of the magnetic field, and the angle at which the coil moves through the magnetic field. The induced voltage is ac and can be brought out of the generator via a pair of slip rings.

Motors and generators are closely related and, on its way to becoming a motor, the unit also works as a generator. The basic difference is in their input. A motor requires an electrical input and supplies mechanical energy at its output. A generator requires mechanical energy at its input and supplies electrical energy at its output.

During the time a motor is functioning, the armature, rotating in the magnetic field supplied by the field coils, develops a voltage that is in opposition to the applied voltage. This counter electromotive force (EMF) limits the armature current to a safe amount.

PHASE

A pair of ac generators can be connected so that their sine-wave outputs follow each other not directly, but separated in time or measured in degrees. Thus, the sine waves from two generators could be separated by 90° (Figure 11–2). A better method would be to have a single ac generator equipped with a pair of armature coils mounted at right angles, with each producing an ac sine-wave voltage output. A pair of such voltages is said to be out of phase. In this case the phase separation is 90°. The first voltage leads the second voltage by this amount. Alternatively, the second voltage lags the first, also by 90°. The generator can be considered as two ac voltage generators combined into one.

This concept of a single generator developing a pair of out-of-phase sine-wave voltages can be carried a step further, as indicated in Figure 11–3. Drawing a shows a single-phase output, while drawing b is that of two-phase ac. Drawing c shows the result of using a single generator with three independent armatures producing three independent ac voltages. Each voltage is referred to as a *phase,* identified as phases A, B, and C, in which the phase angle is 120°.

OUT-OF-PHASE VOLTAGES AND CURRENTS[2]

The output of a single-phase generator consists of a voltage and a current that are in phase. This in-phase relationship continues as long as the load is resistive. This would be the case if the load consisted of a broiler or an electric heater. However, if the load included a coil, the current would lag the voltage, and the separation of

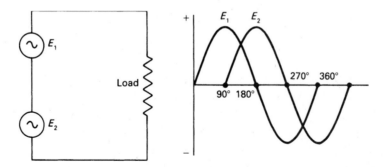

Figure 11–2. The voltage output of two ac generators separated by 90°.

[2](*NEC: Article 455; page 483*)

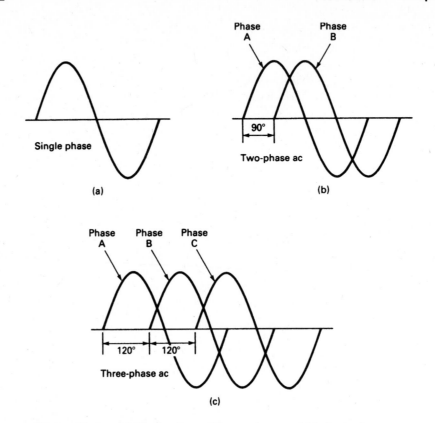

Figure 11–3.   (a) Single-phase, (b) two-phase and (c) three-phase ac.

the two would be a maximum of 90°. If the load included a capacitor instead of a coil, the separation of the current and the voltage would once again be a maximum of 90°, but this time the current would lead the voltage.

THREE-PHASE GENERATOR VOLTAGES

An ac generator can have a single complete armature. However, it can also have three independent armatures with each having an ac output. This generator has three single-phase windings so that the voltage induced in each winding is 120° out of phase with the voltages in the other two windings.

The problem with having three armatures is that each armature coil has a pair of output leads, and since there are three coils, there are six such leads.

Delta Connection

Instead of having the armatures operate independently, they can be connected as shown in Figure 11–4. Each armature has 240 volts induced across it, and there are four voltage outputs. Three of the outputs supply 240 volts. One winding is center tapped, and the center tap is grounded. Each end of the winding is 120 volts with respect to ground. This provides three output voltage lines, with the center wire grounded and 120 volts with respect to ground from the other two lines. The advantage of the delta connection is that it supplies two 120-volt lines and three 240-volt lines.

Wye Connection

Figure 11–5 shows an alternative way of connecting the three armature coils, known as a Y or wye. It supplies three 120-volt outputs and three 208-volt outputs. Each output is obtained from a pair of armature windings instead of a single winding in the delta system. In the wye connection the output voltage is not $2 \times 120 = 240$ volts, but instead it is $1.73 \times 120 = 208$ volts since the two voltages are partially out of phase.

SINGLE-PHASE POWER

While single-phase ac is commonly used in home and industry, single-phase generators are rare. In a single-phase system, only one ac voltage is involved. The power delivered to receptacles in the home is single phase, and most home electrical appliances are designed to work from single-phase ac, but this power is generated by a polyphase source.

The generation of polyphase power is preferable since such power is less

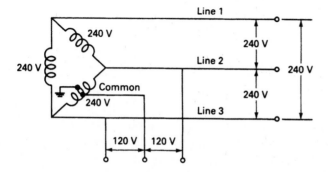

Figure 11–4. Delta line voltages.

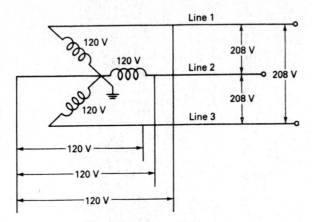

Figure 11–5. Wye (Y) line voltages.

pulsating than single-phase power. Polyphase motors are more efficient, and less copper is needed in a three-wire power distribution line than in a two-wire line. Two phase is the most elementary of the polyphase systems; electric power distribution most often uses three phase.

Each phase of a polyphase generator supplies a voltage that is in phase with the current it supplies. This voltage-current in-phase condition can be shifted externally to the generator depending on whether the load is reactive or not.

POWER AND ENERGY

The words *power* and *energy* are sometimes used synonymously, but incorrectly so. Energy is sometimes chemical, mechanical, thermal, light, kinetic, or electrical. Power is the rate at which energy is being used. Electric power is the rate at which electrical energy is being converted to some other form. An electric current sent through the filament of an electric light bulb is converted into light, mechanical energy when used to drive a motor, or heat when used in an oven or toaster.

You use energy when you lift an object above ground. That object now has potential energy and is capable of falling. Energy is more closely defined by considering its source.

POWER FACTOR

There are three components in ac circuits that have an effect on the amount of current flowing. These are resistance, inductance, and capacitance. Both inductance and capacitance have an opposition to current flow, known as *reactance,* and like resistance it is specified in ohms. The symbol for either inductive or capacitive reactance is the letter X. The combination of resistance and reactance in a circuit is known as *impedance* and is specified by the letter Z. Thus, the impedance in a circuit is one that takes both resistance and reactance into consideration.

Power factor (PF) is the ratio of resistance to impedance and can be written as

$$PF = \frac{R}{Z}$$

The result is a decimal and can be converted to a whole number by multiplying it by 100. Thus, a PF of 0.57, when multiplied by 100, is a power factor of 57%.

AC POWER

In a dc circuit, one formula for the calculation of power is

$$P = E \times I$$

This formula is simple since, in a dc circuit, no frequency is involved. However, in an ac circuit there are two possible reactances, that due to the presence of a capacitor, called capacitive reactance, and that due to the presence of a coil, called inductive reactance. The power in an ac circuit is expressed by

$$P = E \times I \times \frac{R}{Z}$$

But $\frac{R}{Z}$ is the power factor, and so this formula can be written as

$$P = E \times I \times PF$$

REAL POWER

Although electrical power is invisible, it can be regarded as a commodity that is delivered to a consumer. However, in the case of ac the fact that it is delivered does not mean that all of it is accepted and used. Power that is used and not returned to an electric utility is called *real power*. Power that is returned to the source is referred to as *reactive power*.

Since real power involves resistance only, it can be written as

$$P = I^2 R$$

If an ac circuit contains reactance only,

$$P = I^2 X$$

If an ac circuit contains both resistance and reactance, the unused power is referred to as *apparent power*. The ratio of real power to apparent power is the power factor.

Real power, also known as *true* power, is the power actually used and is not power that is returned to its source. All resistive components use real power. A component that not only has resistance, but some capacitance and/or inductance as well, returns some portion of the delivered power back to the source.

Wattmeters measure real power only. Apparent power can be calculated by volt-amperes. Hence, the power factor can also be written as

$$PF = \frac{\text{watts}}{\text{volt-amperes}}$$

VARS

The word *vars* was coined to represent the phrase volt-amperes reactive and consists of the first letter of these words. A *var* is the reactive power of 1 reactive volt-ampere. Here reactive means the presence of a coil or capacitor in a branch power line, volt is the electrical pressure, and reactive amperes refers to the current that is unused and returned to the source.

Since reactive volt-amperes are not measured by a wattmeter, an electric utility receives no payment for them. Consequently, a factory having a high level of reactive power will be billed at a higher rate. This is not applicable to a home since most of the appliances use real power and the typical power factor is high.

Single-phase Power

Single-phase power is stated as

$$P = E \times I$$

The answer is in watts, but this unit is much too small. Instead, power is more conveniently expressed as

$$\text{kilovolt-amperes (kVA)} = \frac{E \times I}{1000}$$

where 1 kilovolt equals 1000 volts.

Three-phase Power

The formula for determining three-phase power is almost the same as that for single-phase power, except for a multiplying factor, 1.732. To determine three-phase kilovolt-amperes,

$$kVA = \frac{E \times I \times 1.732}{1000}$$

THE DC GENERATOR

The voltage produced by a generator is initially AC. In a DC generator, AC voltage is rectified, either mechanically or electronically, appearing as DC at the output. If necessary, DC output can be filtered to supply a smoother output. Quite commonly the DC generator uses a commutator that takes the current output, which always flows in the same direction. This DC generator does not have the nonvarying voltage obtained from a battery. An electronic filter can be used to obtain a more unchanging output. This assumes that the load demands a more truly DC voltage source.

THE MOTOR GENERATOR

A motor generator is often intended for use in industry for producing either an AC or DC voltage. It is needed where a motor requires a substantial DC input not capable of being supplied by batteries or an electronic power supply. It is also used where AC is required but there is no access to a power line, as in the case of a ship.

There are four possible motor generator arrangements:

DC motor driving an AC generator
DC motor driving a DC generator
AC motor driving an AC generator
AC motor driving a DC generator

In the case of the second item listed above, the combination is needed when the available DC voltage is inadequate and must be increased to be able to drive a certain type of motor. Unlike AC, DC voltage cannot be stepped up through the use of a transformer. The drawing in Figure 11–6 shows the basic arrangement of a motor generator.

The drive shaft of the motor can be mechanically linked to a generator by some kind of coupling device or by a gear train. In the case of the arrangement indicated in the drawing, the armature of the motor and that of the generator will rotate at the same speed. A gear train can be used instead so that the generator armature rotates at a higher or lower speed.

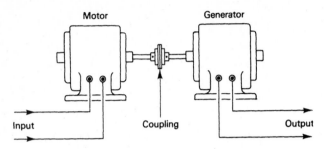

Figure 11–6.   Motor generator.

Advantages and Disadvantages of the Motor Generator

The motor generator supplies simple output voltage control. As a source voltage it has excellent regulation.

Regulation.   The output voltage of a motor generator is maximum under conditions of no load. As the load increases, the output voltage will possibly become lower. Regulation means the amount of voltage decreases, when a full load is applied, compared to the voltage with no load.

As an example, suppose the output voltage is 600 with no load, but drops to 510 with full load. The no-load voltage of 600 minus the full-load voltage of 510 is 90 volts. Dividing this by the no-load voltage is 90 divided by 600, or 0.15. The regulation is expressed as a percentage, thus a regulation of 15%. A regulation of about 10% is considered good.

In terms of a formula, this can be expressed as:

$$\text{Percent regulation} = \frac{100(E1 - E2)}{E2}$$

where $E1$ is the no-load voltage and $E2$ is the full-load voltage.

The motor generator does not supply the pure DC output obtained from batteries and has some ripple (a variation in the DC output).

The motor generator is large and heavy, and is usually more expensive than electronic supplies or the equivalent in batteries. However, it does have the advantage that it can be used where batteries would be impractical and no AC source is available.

The motor generator is heavy and noisy. It can cause interference to radio receivers. Repairing the motor generator can be difficult. It requires a regular maintenance program that includes lubrication and replacement of brushes.

THE ROTARY CONVERTER

The rotary converter shown in Figure 11–7 resembles the dynamotor as far as the field coils and commutator are concerned. The output is AC and is obtained by a pair of slip rings. Unlike a dynamotor the rotary converter has just a single armature, which works for both the generator and motor sections of the unit.

Because of its construction, the function of the converter can be transposed, that is, an AC voltage can be applied to the slip-ring output and DC can be obtained from the commutator.

Since the input and output are so closely related, the output voltage can fluctuate with voltage changes at the input. When it works as an AC motor it is of the synchronous type, and in this mode is sometimes referred to as a synchronous converter.

THE DYNAMOTOR

The dynamotor, shown in Figure 11–8, is a combined motor-generator. It is a DC-to-DC machine having a DC voltage input and a DC voltage output. Its input is low-voltage DC, such as that supplied by a storage battery. Its generator output is high-voltage DC. Thus, it does for DC the equivalent function that transformers do for AC. The transformer has an advantage since it has no moving parts.

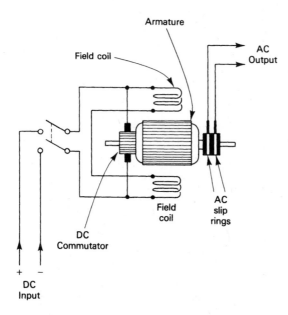

Figure 11–7.   Rotary converter.

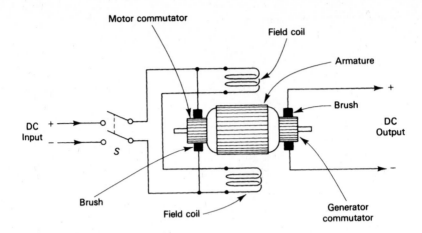

Figure 11–8. Dynamotor.

The drawing of the dynamotor shows that the unit has an armature and field coils that are common to the motor and generator sections of the unit. Although the drawing shows a single armature there are two independent windings with each occupying slots in the armature core.

The unit is equipped with two commutators, one for the DC input and the other for the generator section. The motor commutator permits the battery source supplying DC input to give current to the rotating armature. The generator commutator works as a rectifier converting the AC supplied by the armature to DC. In both instances for the motor and generator commutators, input and output currents are supplied via brushes.

The DC-to-DC dynamotor is used in industry where high voltage and high direct current are required. There are also AC dynamotors that have a DC input and an AC output. The input to the motor is the same as in the DC-to-DC dynamotor. However, the AC output does not have a commutator. Instead, the alternating current supplied by the section is brought out to a pair of slip rings. These also use brushes for the delivery of alternating current to an AC load. The DC-to-AC dynamotor is used in transportation where no source of AC is available.

SUMMARY OF FORMULAS FOR GENERATORS

Generator frequency:

$$f = \frac{\text{rpm} \times p}{120}$$

where f is the frequency in hertz, p the number of poles, and rpm the number of revolutions per minute.

Rotating speed of a generator:

$$\text{rpm} = \frac{120 \times f}{p}$$

Number of poles of a generator:

$$p = \frac{120 \times f}{\text{rpm}}$$

Star-wound AC generator:

$$E_{\text{line}} = 1.732 \times E_{\text{two-phase}}$$

Power in a single-phase reactive unit:

$$P = E \times I \times \text{PF}$$

where P is the power in watts, E is the voltage in volts, I the current in amperes, and PF the power factor.

Power in a three-phase reactive unit:

$$P = E \times I \times \text{PF} \times 1.732$$

Current in a three-phase reactive unit:

$$I = \frac{P}{E \times \text{PF} \times 1.732}$$

Voltage in a three-phase reactive unit:

$$E = \frac{P}{I \times \text{PF} \times 1.732}$$

Single-phase current:

$$I = \frac{746 \times \text{hp}}{E \times \text{eff} \times \text{PF}}$$

where eff is the efficiency.

Three-phase current:

$$I = \frac{746 \times \text{hp}}{E \times \text{eff} \times \text{PF} \times 1.732}$$

where hp is horsepower.

Single-phase kVa:

$$kVa = \frac{E \times I}{1000}$$

where kVa is kilovolt-amperes.

MOTOR TYPES AND CHARACTERISTICS

Horsepower

The word *horsepower,* abbreviated hp, was originally introduced in connection with steam engines. It is often used with reference to both dc and ac motors and is an indication of the ability of the armatures of these motors to do work and the rate at which that work is done. One horsepower is equivalent to lifting a weight of 33,000 pounds a distance of 1 foot in 1 minute. Or 1 hp = 33,000 ft-lb (foot pounds) per minute. A foot pound is exactly that, moving 1 pound a distance of 1 foot.

Since there are 60 seconds in 1 minute, divide 33,000 by 60, and the result is 550. One horsepower, then, is also equal to 550 ft-lb per second. Stated in terms of a formula:

$$hp = \frac{\text{number of ft-lb per minute}}{33,000}$$

Electrical machinery is rated in either horsepower or watts. In a formula,

$$1 \text{ hp} = 746 \text{ watts or } 1 \text{ watt} = 0.001341 \text{ hp}$$

To change horsepower to watts, multiply horsepower by 746. Since 1 hp = 746 watts and 746 watts = 0.746 kilowatts (kW), it is possible to convert directly from horsepower to kilowatts by multiplying horsepower by 0.746. The 746 figure is an approximation and a more nearly correct value is 745.7. However, 746 is practical for electrical problems.

Horsepower and watts are two different ways of indicating rate of work or power, not quantity. Motors are often rated in terms of horsepower, but this rating indicates the output capability, not the input. A 1-hp motor could do work equivalent to 746 watts, but to get this amount of output, the input power would have to be about 20% greater to allow for various losses in the motor.

Mechanical Power versus Electrical Power

Mechanical power is a reference to horsepower, while electrical power is represented by a basic unit, the watt (W). The letter P is used as an abbreviation for power. One formula for electrical power is

$$P = E \times I$$

where P is the power in watts, E is the voltage in volts, and I is the current in amperes. The relationship between mechanical and electrical power can be expressed as

$$\text{hp} = \frac{\text{volts} \times \text{amperes}}{746}$$

or

$$\text{hp} = \frac{E \times I}{746}$$

To determine the horsepower equivalent of electrical power in watts, divide watts by 746. If the power is in kilowatts (thousands of watts) or megawatts (millions of watts), first convert to watts before substituting in the formula.

Horsepower is a basic unit of measurement, like the ohm (the basic unit of resistance), the volt (the basic unit of electrical pressure), and the ampere (the basic unit of current). Horsepower is used for dc and ac motors and also for electrical generators. The input to a motor is electrical and is expressed in watts or in some multiple of the watt, such as the kilowatt (kW). The output of a motor, its mechanical power, represents the turning force of its armature and is stated in horsepower.

Both dc and ac motors are commonly designated in terms of horsepower, and this is a very wide range. Fractional-horsepower motors could be rated as $\frac{1}{100}$ or even less, while typical designations could be $\frac{1}{4}$, $\frac{1}{3}$, $\frac{1}{2}$, and $\frac{3}{4}$, integral sizes could be a whole number plus a fraction. Typical values could be 1, $1\frac{1}{4}$, $1\frac{1}{2}$, $1\frac{3}{4}$, 2, 3, 5, $7\frac{1}{2}$, 10, 15, 20, and so on.

Time is not involved in these designations. Whether it works for a minute or an hour, a 1-hp motor delivers 1 hp for as long as it operates.

The Overloaded Motor[3]

A motor rated at $\frac{1}{2}$ hp can be expected to supply this amount of mechanical power for as long as the motor operates, and it has this rated capability whether its load is variable or fixed. A varying load also presents the possibility of having a higher than rated load, that is, overloading the motor. Usually, a motor can tolerate a 25% overload, and motors are often designed to have this capability. There is a time element involved, and the overload should not exceed the rated operating load for more than 2 hours.

[3](*NEC: Table 430–37; page 432*)

Variable Electrical Power Input

The greater the loading on the shaft of a dc or ac motor, the greater will be the demand for electrical power input. If there is a 10% increase in load, there will be at least a 10% increase in the demand for electrical power input, but this assumes the efficiency of the motor remains constant. If motor friction increases or if its heat losses rise more than usual, a 10% increase in load may mean an increase in electrical input power demand of 15%, the exact amount depending on the design and structure of the motor.

TORQUE[4]

Torque is the force required to obtain rotation of the armature; it is the turning power of either a dc or ac motor. The torque of a motor is expressed in a number of ways, and these must be stated before assigning numbers. Torque is stated in inch-ounces (in-oz) for small motors and in foot-pounds (ft-lb) for larger ones.

Inch-ounce and Foot-pound

The inch-ounce and the foot-pound are both units of work. The names supplied here are descriptive. An inch-ounce is the work done by a force of 1 ounce working through a distance of 1 inch. A foot-pound is the work done by a force of 1 pound working through a distance of 1 foot. Note in these statements that no time units are involved. Lift a suitcase weighing 10 pounds and carry it for a distance of 30 feet and the work done is 10×30 or 300 foot-pounds. This remains the same whether the work requires 5 seconds or 5 minutes. However, this example is somewhat simplified since the weight of the person is not included nor is the distance through which the suitcase is lifted. A person weighing 120 pounds walking a distance of 30 feet does $120 \times 30 = 3600$ foot-pounds of work. This must be added to the figure of 300 foot-pounds obtained previously.

Types of Torque

There are a number of ways of describing torque, depending on what a motor is doing at the moment.

Full-load torque.   This is the torque of a motor working under load. If the load is constant, one torque figure will be obtained, but if it is variable, there

[4](*NEC: Table 430–6b; page 417*)

will be a succession of torque figures. The full-load torque for a motor under constant load conditions is

$$\text{full-load torque} \ = \ 5252 \ \times \ \frac{\text{horsepower}}{\text{rpm}} \ \ (\text{in ft-lb})$$

Starting torque.   This could also be referred to as the maximum torque since the motor must overcome the inertia of both the armature and the load. It is at this time that the input electrical power is maximum.

Rated starting torque.   This specification is supplied by the manufacturer of a motor. It is the minimum torque produced by a motor geared for 1 rpm at its rated voltage whose frequency is 60 hertz, working at an ambient temperature of 25°C when energized through a snap-type switch. The motor may not reach its rated speed while developing its rated starting torque.

Pull-up torque.   An armature will start to rotate when an electric current begins to flow through its windings. During this time the armature will accelerate its rotation until it reaches its operating speed. There will be an increase in torque between the starting speed and operating speed, called pull-up torque.

Torque Requirements

A motor may be part of some electrical device, such as a fan, vacuum cleaner, or refrigerator. Thus, the torque of the motor is the concern of the manufacturer of the electrical component. It is of interest when the component must be repaired. However, there are some instances when an independent decision is required about the torque of a motor. The motor may be needed to operate a machine tool, such as a lathe, grinder, or stand-alone sander. The motor must have enough torque to meet the requirements of the load.

SOFT-START MOTORS

A motor that always starts with a light load with more than enough torque to handle a heavier load is sometimes referred to as a soft-start type. Such motors include the shaded-pole induction motor, the split-phase, and the permanent split-capacitor induction types.

In some instances a motor will work into a load that ranges from extremely light to heavy, as in the case of a motor that is part of an electric drill. The torque demand depends on the kind of material being drilled and the size, sharpness, and material of the drill bit.

RATED GEAR TRAIN LOAD[5]

The mechanical output of a motor may work directly into a load or indirectly via a gear train, belt, or pulley. The rated gear train load is the amount of torque available from that train. There are two torque ratings: continuous duty and intermittent duty. Standard gear trains are rated at 20 inch-ounces continuous duty or at 80-inch ounces intermittent duty. These ratings are independent of motor speed. It is inadvisable to exceed these ratings.

MOTOR SPEED

Motors are sometimes characterized by the rotational speed of the armature, and is specified in rpm, but these are always maximum values. For miniature motors the shaft speed is about 10,000 rpm, and it is about 5000 rpm for small motors rated at $\frac{1}{2}$ hp or less. For some motors the maximum available speed depends on the frequency of the ac source. For induction motors the maximum shaft speed is 3450 rpm when the line frequency is 60 hertz, but it decreases to 2875 rpm for 50 hertz. The selected line frequency of a power utility has an extremely small variation, as evidenced by the accuracy of electric clocks.

In some instances the line input to a motor will be supplied by an independent ac generator. The frequency of the output is often 400 hertz, and for these a wide range of maximum motor speeds is available. Typical rpm values are 3700, 5400, 7200, 10,500, and 21,000. For constant-speed motors, representative values are 1140, 1725, and 3450 rpm.

EFFICIENCY

No electrical machine is 100% efficient. There are always some energy losses in going from one form of energy to another. In the case of a motor the input is electrical energy; the output is mechanical energy.

The efficiency of a machine is expressed as a percentage. A machine rated at 50% efficiency loses half the energy put into it. No energy is ever lost, and so "losses" in this case means the energy is changed to a form that cannot be utilized. Most often the lost energy is in the form of a low level of heat.

Efficiency expressed as a decimal instead of a percentage is known as the

[5]*(NEC: Article 100; page 31)*

efficiency factor. An efficiency of 65% is the same as an efficiency factor of 0.65. In terms of a formula,

$$\text{output} = \text{efficiency} \times \text{input}$$

To find the output of a machine, multiply the input by the efficiency factor. For example, if a motor has an input of 800 watts and an efficiency factor of 85%, the effective output in terms of horsepower can be calculated as

$$\text{output} = 0.85 \times 800 = 680 \text{ watts}$$

However, the output of a motor is in terms of horsepower, not in watts: 1 HP = 746 watts. $\frac{680}{746} = 0.91$ hp. The difference between 1 hp and 0.91 hp is $1 - 0.91 = 0.09$ hp, and this amount of energy is lost in the form of heat.

The efficiency factor can be combined with formulas involving either power in terms of watts or horsepower. The formula for horsepower in terms of the voltage E, in volts, and the current I, in amperes, is

$$\text{hp} = \frac{E \times I}{746}$$

Transposing the terms in this formula,

$$746 \times \text{hp} = E \times I$$

This formula can be rearranged to read

$$I = \frac{746 \times \text{hp}}{E}$$

But this does not take the efficiency into account. If the efficiency is less than 100%, and it always is,

$$I = \frac{746 \times \text{hp}}{E \times \text{eff}}$$

Efficiency and current have an inverse ratio. If the efficiency increases, the current requirement of the motor decreases.

LINE VOLTAGE FOR MOTORS

Fractional-horsepower motors, such as those used in the home, operate from a line voltage of 120 volts. However, heavy-load devices such as washers and dryers operate from 240 volts. Motors for industrial use require 480 volts, while those that are rated at 100 hp or more may have ac inputs of 2400 volts.

DEDICATED BRANCH LINE[6]

Most motors used in the home, other than those intended for workshop use, are limited horsepower types and operate from a single-phase 120-volt branch line. For high-power motors, a separate single-phase branch may be installed, possibly for 240-volt use. This branch may be a line intended for motor use only and in that case is referred to as a dedicated line. If a high-horsepower motor is connected to a receptacle servicing other electrical components such as a radio, TV set, or lamp, turning the motor on may cause them to become momentarily inoperative.

MOTOR TYPES[7]

Universal Motors

The universal motor is so called since it can accept either dc or ac as its input voltage source. Actually, as far as electrical power in the home is concerned, the fact that a universal motor can operate from dc is of little consequence. In the home the universal motor is used to operate appliances such as sewing machines, vacuum cleaners, fans, electric razors, mixers, food grinders, and so on.

Figure 11–9a is the circuit diagram for this motor, while drawing b shows how the motor is assembled. The field coil and the armature coil are in series, and so the same current flows through both. The armature is connected to a commutator. Current flows from the source through a pair of brushes that rest on the commutator. Each pair of commutator bars or segments is connected to an armature winding, so, as the armature rotates, current can flow into one coil, and then another coil, and then still another coil. The electromagnetic field accompanying the armature current reacts magnetically with the magnetic field produced by the field coil. This reaction results in magnetic repulsion, which turns the armature.

There are several different types of universal motors. One of these, known as the compensated series-wound universal, has the field coils positioned in slots. This results in better speed regulation and higher starting torque. Another type of universal motor, known as the straight series, is simpler in construction and is used in sizes that range up to $\frac{1}{3}$ hp, while the compensated type is usually found in motors having a higher horsepower rating.

[6]*(NEC: Article 100; page 24)*
[7]*(NEC: Article 430–7; page 417)*

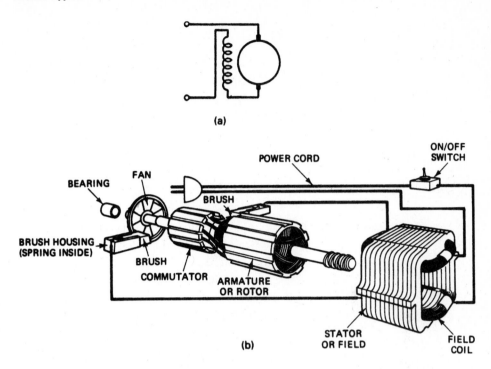

Figure 11–9. Universal motor. (a) Diagram (b) assembly details.

The universal motor has a higher average operating speed than many other types of ac motors and, further, this speed can be controlled by a resistor inserted in series with its line voltage. The problem with this arrangement is that the full line current must flow through the resistor, generating heat and representing an uneconomical way of adjusting the speed. The control can be continuously variable, a form preferred in certain tools such as electric drills. The speed control can be a tapped inductor type, which permits the quick selection of previously predetermined speeds. The inductor is actually part of the field winding and increases or decreases its magnetic field. This speed-control method produces less heat than the resistor control and is also more economical in operation.

A disadvantage of both types of speed control is that neither can control the direction of rotation of the armature. Reversing the line plug will have no effect. A change in direction can be obtained by transposing the connections to either the field or armature windings, but not both. However, the universal motor operates more efficiently in one direction than the other, with the preferred direction being selected by the manufacturer of the motor-operated appliance.

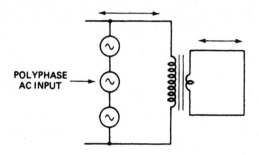

POLYPHASE
AC INPUT →

Figure 11–10. Armature of squirrel-cage motor is equivalent to shorted turn secondary of transformer.

SQUIRREL-CAGE MOTORS

The strength of the magnetic fields surrounding field coils and armature coils depends on the inductance of these coils. The inductance can be increased by using more coil turns and by using a metallic core. The core is not solid, but is constructed of thin laminations of sheet steel. In effect, this motor is somewhat like a transformer, except that the armature, comparable to the secondary winding of the transformer, is capable of rotation. In a transformer the secondary winding, as shown in Figure 11–10, has a greater flow of current as the number of its turns is decreased. The maximum current is a shorted secondary. The drawing shows a single-turn secondary, equivalent to a maximum turns ratio.

    The squirrel-cage motor takes advantage of this step-down transformer

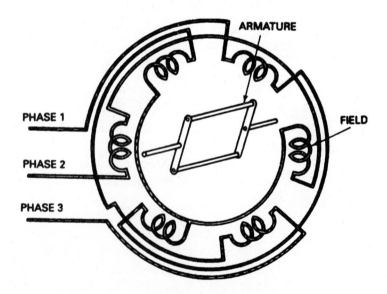

Figure 11–11. Field and armature of squirrel-cage motor.

characteristic and works by electromagnetic induction, producing a large flow of current in an armature known as a squirrel cage. In the transformer the electrical energy is supplied by the primary winding. In the squirrel-cage motor the energy is supplied by the field coils.

This motor, also known as an induction motor, does not have a commutator. The line voltage is connected directly to the field coils, but this does not present a problem since these coils remain fixed in position. The armature has no connections. It obtains its magnetic field purely by induction. Figure 11–11 shows the basic structure.

Figure 11–12 shows two types of armatures. Drawing a uses copper strips

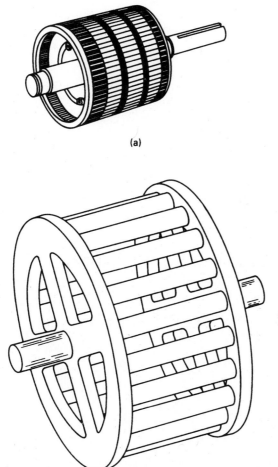

(a)

(b)

Figure 11–12. (a) Copper strip armature and (b) copper rod armature.

connected to circular end pieces, with the entire assembly mounted on a shaft. Drawing b shows the use of copper rods instead of strips, but other than this physical difference both armatures work in the same way.

Motor Speed

In a squirrel-cage motor, there are a number of independent field coils, with each becoming an electromagnet when a line current flows through them. Each of these magnets has a north and a south pole, with the speed of the armature determined by the number of pairs of poles. The speed is also governed by the frequency of the input voltage, but most often this is a nonvariable 60 hertz. Figure 11–13a shows the symbol.

Figure 11–13b is an exploded view of this motor. It is often a polyphase voltage input type using a wye structure for the field coil, although a delta arrangement is also used.

(a)

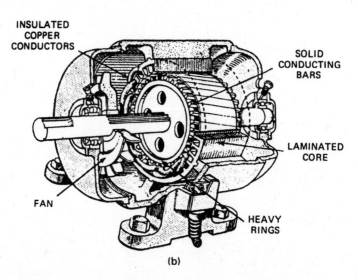

(b)

Figure 11–13. (a) Squirrel-cage motor symbol; (b) exploded view.

Characteristics

The squirrel-cage motor can have a high starting torque, which is necessary for heavy loads and is used in applications such as hoists and heavy doors. It is noted for the simplicity of its construction and its good efficiency.

Because of its three-phase ac input requirement, this motor is not suitable for in-home use since in-home voltage is single phase. Typically, a three-phase input is for motors rated at about 3 hp, but this requirement can be met by single-phase motors as well. The squirrel-cage motor draws a lagging current, a characteristic that is to be expected in a motor using coils having a high inductance.

INDUCTION MOTORS[8]

The squirrel-cage motor can be described as a repulsion type or as an inductor type. It can be called repulsion because of the force of repulsion between the magnetic poles of the field coils and the magnetic poles of the armature. It can be called induction because the magnetism around the field coils induces an opposing magnetic field around the armature.

There are repulsion-induction motors other than the squirrel-cage motor, but these are variations of the basic squirrel cage. These motors include the wound rotor induction, repulsion start induction, shaded pole, hysteresis synchronous, hybrid, split-phase induction, and capacitor-start, including the two-speed capacitor-start motor, the permanent capacitor motor, and others.

REPULSION-INDUCTION MOTORS

The repulsion-induction motor (Figure 11–14) is a variation of the wound rotor and the squirrel cage since it has two armature arrangements. One of these is a squirrel cage, while the other uses wound coils. Like other motors that use a wound armature, the motor requires a commutator with brushes delivering a current via the commutator to the armature.

The armature, then, is a hybrid type, but there are certain advantages to this arrangement. In its operation it works somewhat like a dc motor. It has a high starting torque and its operating speed remains fairly constant. It is also possible to change the direction of rotation of the armature by shifting the brushes by

[8](*NEC: Article 430–3; page 414*)

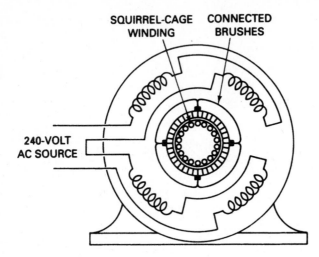

Figure 11–14. Repulsion-induction motor.

about 15°. The brushes are mounted in a movable brush holder, and so making the change is very simple. Also, as a guide to making an armature rotation change the amount of brush shift and the direction of the armature are marked either on the brush holder or on the motor frame. As a further aid, there may be limit stops to help in a more accurate way of moving the brushes.

A further modification found in the repulsion-induction motor is a compensating winding, shown in Figure 11–15, to aid in improving the motor's power factor. This could be a help in an industrial situation where power factor is important in keeping electrical energy costs as low as possible.

Some repulsion-induction motors are fractional-horsepower types, ranging

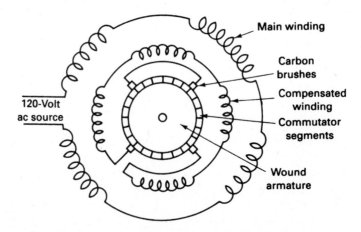

Figure 11–15. Compensating windings in repulsion-induction motor.

from $\frac{1}{8}$ to $\frac{3}{4}$ hp. The current rating extends from 4 amperes for the $\frac{1}{8}$-hp motor to 15 amperes for the $\frac{3}{4}$-hp machine.

SHADED-POLE MOTORS

The shaded-pole (also known as interpole) motor (Figure 11–16) is also a modified squirrel-cage type. It looks just like a squirrel-cage unit except that it is equipped with a pair of shading coils. While just a single pair of these coils is shown, there could be more. Their purpose is to make the motor a self-starting type; hence, they are popularly used in electric clocks.

   This motor uses a squirrel-cage rotor and has no need for a commutator and brushes. To produce a shaded pole, the laminations of the field winding are notched and then fitted with a single heavy ring of copper. The line cord of the motor is connected directly to a single-phase, 120-volt receptacle.

Characteristics

This motor is somewhat inefficient and has a low starting torque, but this latter characteristic is not too important since the load, as in the case of a clock, is very light. Shaded-pole motors are fractional-horsepower types, and $\frac{1}{25}$-hp units are common. The rotational speed of the armature is 900, 1800, or 3600 rpm. Rotation cannot be reversed. They are found not only in clocks, but in motor-operated toys, hair dryers, small fans, and advertising displays. Speed control can be obtained by using a resistor or an iron-core inductor in series with the line voltage.

   Another type of self-starting motor has a nonlaminated rotor made of steel, which becomes a permanent magnet. The stator has a rotating magnetic field, and it forces the turning action of the rotor. The motor is equipped with a gear train that reduces the rotation of the shaft to 1 rpm. In a clock, this is used for the

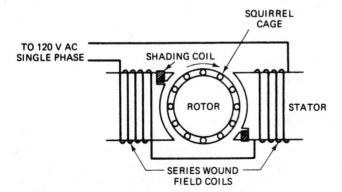

Figure 11–16. Shaded-pole motor.

sweep second hand, while another gear train is used for the minute and the hour hands.

SYNCHRONOUS MOTORS

The synchronous motor is a constant-speed type. The armature turns in step with the rotating magnetic field of its field coil. This can be the power-line frequency or some multiple.

The speed of rotation of the armature is set by the number of pairs of poles and the frequency of the ac source voltage. In terms of a formula, the speed is

$$\text{rpm} = \frac{f \times 60}{p}$$

where rpm is the rotation of the armature, f is the frequency of the source voltage, and p is the number of poles (p is an even number such as 2, 4, or 6 since the poles are in pairs).

Synchronous motors of the type shown in Figure 11–17 are commonly used in clocks. The motor has a close resemblance to transformer structure since it consists of hardened steel shaped in a set of E and I laminations, so called because of their shape resembling these letters. The line voltage is supplied to the I laminations of a field coil. The armature is a squirrel-cage type and, as shown, the motor is equipped with shading poles.

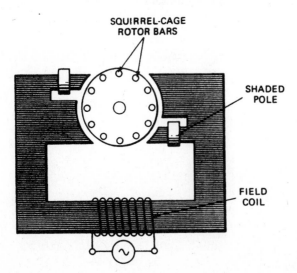

Figure 11–17. Synchronous clock motor.

Characteristics

The speed of this motor is controlled by the frequency of the line voltage. This is remarkably constant, and over a 24-hour period the frequency shift is zero. The motor is self-starting, and its speed is constant whether loaded or unloaded. However, it does have a tendency to decrease its speed with an increase in the load.

SELF-STARTING INDUCTION-REACTION SYNCHRONOUS MOTORS

Some motors are made to order for specific applications and so may never be listed in manufacturers' catalogs. These are based on existing motor types and may be made to emphasize certain wanted motor characteristics. In some cases the motor may be a combination of induction and synchronous motors and may be equipped with six shaded poles, also known as *salient* poles. This particular motor uses a squirrel-cage rotor.

While motors are ordinarily designed for line voltage input, some are intended for use with a separate ac generator. The generator may be a fixed-frequency type whose output is some frequency other than 60 hertz. There are also some motors intended to be driven by a number of different frequencies supplied by a variable-frequency oscillator (VFO). The problem with external oscillators is that they may not be able to supply enough current for the field coils even though the motors are small, fractional-horsepower units. In this case the oscillator is followed by a power amplifier that can meet the current requirements of the driven motor.

COMBINED AC/DC MOTORS

Ordinarily, the electrical power input to a motor is either ac or dc, but a motor known as polyphase synchronous requires both. As shown in Figure 11–18, the stator windings are supplied with a three-phase input, while the rotating coils get their input from a dc supply. In series with the dc supply is a variable wirewound resistor, R, working as a speed control. The dc supply can be either a battery or an electronic power supply obtaining its input from the ac line. The dc supply is connected to a pair of slip rings that feed current to the field coils connected in series. Unlike other motors, it is the field coils that rotate, supplying a rotating magnetic field. In some motors the field coils are permanent magnets, thus eliminating the need for a separate dc supply. The fixed position coils, known as the stator, are the coils receiving the three-phase ac input.

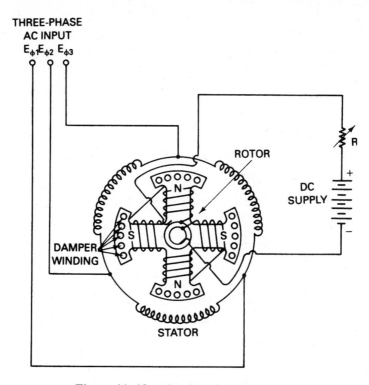

Figure 11–18.   Combined ac–dc motor.

HYSTERESIS SYNCHRONOUS MOTORS

The copper rods or strips of squirrel-cage motors become temporary magnets during the time that they have induced currents from the field coils. This induced current is accompanied by a magnetic field, and this is the case with all current-carrying conductors. When the current stops, the magnetic field collapses and disappears.

If the squirrel-cage conductor consists of a material such as iron, steel, or an alloy that includes cobalt, these conductors also have a magnetic field when current flows through them. When the current stops, the magnetic field does not disappear completely; some of it remains and is known as *residual* magnetism.

The hysteresis synchronous motor takes advantage of this characteristic and is constructed as shown in Figure 11–19. This motor is a fractional-horsepower type, and its field coil gets its current from a single-phase ac line. The motor has a pair of shading coils; hence it is self-starting. The magnetic field accompanying the field current varies in frequency and so is a rotating type. This magnetic field cuts across the rotor from the alternating poles of the field laminations. Although only two poles are shown in the drawing, the motor can be equipped with more. The rotational speed of the rotor is determined by the number of poles and the fre-

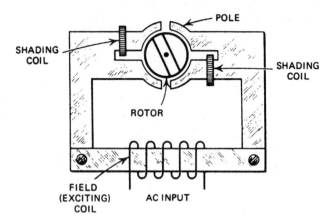

Figure 11–19. Two-pole hysteresis synchronous motor.

quency of the line current. If the input frequency is 60 hertz and the motor has just one pair of poles, the speed of the rotor will be 1800 rpm.

SPLIT-PHASE INDUCTION MOTORS

The split-phase induction motor (Figure 11–20) is a member of the squirrel-cage rotor family. It operates from a single-phase, 120-volt ac line and is equipped with field coils, including a starting winding and a running winding.

While the ac input is a single-phase line, the current it supplies can be made into a pair of out-of-phase currents. As a result, the motor behaves as though connected to a two-phase source, that is, as a polyphase motor. Out-of-phase currents can be obtained by two controlling factors: (1) the inductance of the coil and (2) its resistance. This can be accomplished by having a pair of coils connected in series. One coil could be made to have a large amount of inductance compared to its resistance. In this case, the phase of the coil would be close to 90°. This could be achieved by using a larger number of turns of low-resistance wire. The second coil could have a smaller number of turns of high-resistance wire. In this case, the flow of current through it would have a phase angle somewhat closer to 0°.

The motor is equipped with a pair of switches, a main switch and a starting switch. The main switch can be operated manually, while the starting switch, a centrifugal type, is initially in its closed position. After the motor has reached about 75% of its running speed, the centrifugal switch opens.

This motor is a fractional-horsepower type and has moderately good speed regulation. Horsepower ratings are from $\frac{1}{50}$ hp, corresponding to 14.9 watts, to $\frac{1}{3}$ hp, equivalent to 248.66 watts. It has low starting torque, just a fraction of its torque at full running speed. The motor is intended for easy-to-start light loads. The starting current is high and can be as much as six to eight times the full-speed

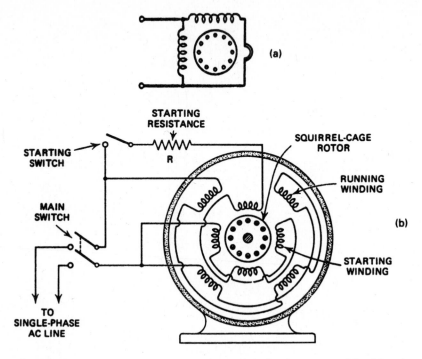

Figure 11–20.   (a) Symbol for split-phase induction motor; (b) wiring arrangement.

running current. The motor is used for oil burners, office appliances, fans and blowers, and light load appliances.

Centrifugal Switches

As a motor rotates it develops an outgoing force called centrifugal. This force can be used to operate switches; some of these are illustrated in Figure 11–21. The two that are shown are (a) the ballweight type and (b) the flyweight.

The switch in drawing a has a pair of ball weights kept in an initial position by a spring. Before the armature turns, the spring holds the two contacts together, and so there is a complete current-conducting path to voltage for the starting winding. After the armature is in motion, centrifugal force drives the two ball weights outward. This opens the current connection to the starting winding, and so that winding is no longer effective. When the motor stops, the centrifugal force stops, and the ball weights are pulled inward by the spring, closing the circuit for the starting winding.

Drawing b shows the use of flyweights, which are moved outward by centrifugal force. This opens a pair of contacts, thus preventing further current flow to the start winding. The contacts close when the motor stops running.

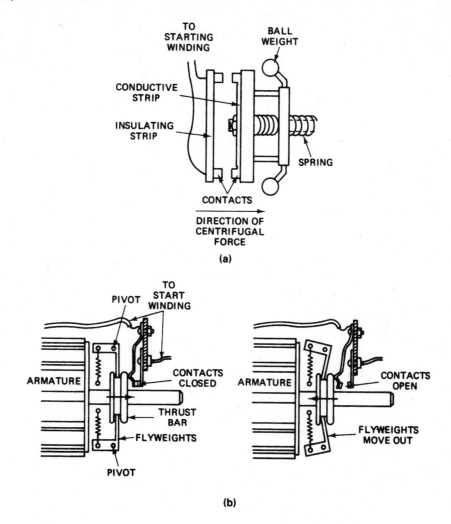

Figure 11–21. Centrifugal switches: (a) ballweight type; (b) fly-weight type.

CAPACITOR MOTORS[9]

Consider a pair of resistors in series with a current flowing through them. These currents produce a voltage across the resistors known as *IR drops* ($E = I \times R$). The currents and the voltages are in phase; that is, the phase angles between them are zero. Substitute a capacitor for one of the resistors and the effect will be to shift

[9]*(NEC: Article 460; page 486)*

the phase of the voltage across it by a maximum of 90° compared to the voltage across the resistor. The result will be the equivalent of a two-phase system. This concept is used by a number of motors, such as:

Capacitor start

Split capacitor

Two capacitor

CAPACITOR-START MOTORS

Figure 11–22 shows the arrangement of a capacitor-start motor. The source is 120 volts ac, single phase. The motor is a squirrel-cage type and is equipped with an auxiliary winding working as a motor start, identified as L1. A second coil, L2, is the main or running winding. Connected to the motor circuit is a starting capacitor, C, which can make contact with a centrifugal switch.

This motor is a split-phase type, with the phase shift obtained by having a capacitor in series with an auxiliary (starting) winding. The capacitor is an oil-impregnated, sealed paper unit.

The motor is started by closing the double-pole, single-throw (DPST) switch. The starting capacitor, the centrifugal switch, and the auxiliary winding are all in series, with this network connected to the 120-volt ac line. When the motor approaches its operating speed, the centrifugal switch opens and the starting capacitor and auxiliary winding are disconnected from the line voltage.

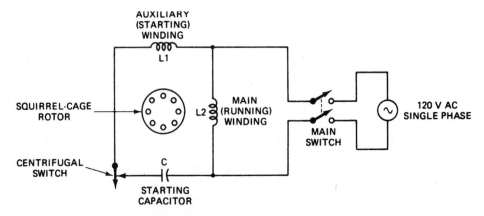

Figure 11–22.   Capacitor-start motor.

Characteristics

Typical units have a horsepower output ranging from as little as $\frac{1}{6}$ hp to as much as 10 hp. They operate from a single-phase 120-volt ac power line, but some units are designed to use 240 volts ac. Full operating speed can be 900, 1200, 1800, and 3600 rpm. Rotation of the squirrel-cage rotor can be reversed. The motor is commonly used for machine tools, grinders, blowers, and ventilating fans.

TWO-SPEED CAPACITOR-START MOTORS

Figure 11–23 shows the arrangement of a two-speed capacitor-start motor capable of operating from either 120 or 240 volts ac, single phase. The unit is equipped with a single-pole, double-throw switch to permit selection of a low or high speed. The motor has a pair of running windings identified as 1 and 2. The centrifugal switch has two operating positions, R (running) and S (start). When the motor starts, the capacitor is included in the working circuit but is disconnected by the centrifugal switch when the motor approaches its running speed. The motor is used for operating two-speed fans.

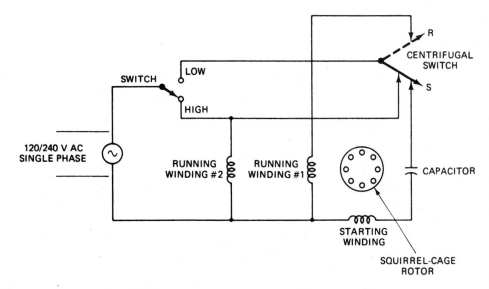

Figure 11–23. Two-speed capacitor-start motor.

TWO-CAPACITOR MOTORS

The two-capacitor motor appears in Figure 11–24. Like the other capacitor motors, it has a squirrel-cage rotor. Its input is 120 volts ac, single phase and is equipped with two stator windings: a main and an auxiliary winding. In series with these windings is a pair of capacitors, one of which can be taken out of the circuit by a centrifugally operated switch. The value of capacitance for the two capacitors is carefully selected so as to obtain the best starting and running torque. When the two capacitors are in shunt, the total capacitance is the sum of the individual capacitances. The parallel arrangement exists when the motor is

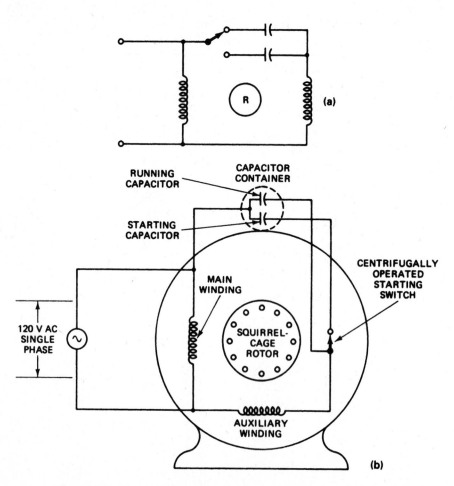

Figure 11–24.   (a) Symbol for two-capacitor motor; (b) wiring arrangement.

first turned on. When the armature is close to its operating speed, the centrifugal switch opens and one of the capacitors is removed.

The two capacitors are different types. The running capacitor, the unit that always remains connected, is a sealed-paper type, while the starting capacitor is an ac electrolytic.

PERMANENT CAPACITOR MOTORS

Sometimes the necessary phase shift can be handled by a single capacitor, so the motor circuitry is simplified since one capacitor and the centrifugal switch are eliminated. While the torque is not as good as in the two-capacitor motor, this motor can be used when the load is constant.

AUTOTRANSFORMER MOTORS

The autotransformer motor is yet another squirrel-cage motor, but its phase shift is obtained by a transformer in parallel with a capacitor. The transformer that is used is an autotransformer. Unlike the usual transformer, in which the primary and secondary windings are physically separate, the autotransformer consists of a single tapped winding. The autotransformer, an iron-core type, generally has a single tap, but can have two to permit its use for two functions, start and run, selected by a single-pole, double-throw switch, called a transfer switch.

The purpose of the autotransformer is to step up the voltage applied to the capacitor. The effect is equivalent to increasing the capacitance, but the capacitor must have its rated working voltage increased accordingly. It also controls the amount of phase shift supplied by the capacitor, since the inductive reactance is in opposition to the capacitive reactance.

Direction of Rotation

Some motors are designed to permit changing the direction of rotation. Whether a motor can reverse depends on both motor design and the way the motor is coupled to its load and is most available when direct coupling is used. Motor reversal, for example, is a common feature of ceiling fans.

Figure 11–25 shows a number of motors and the techniques for reversal. Drawing a is a single-phase capacitor motor equipped with a single-pole, double-throw reversal switch. The purpose of this switch is to transfer the connections from one stator winding to another.

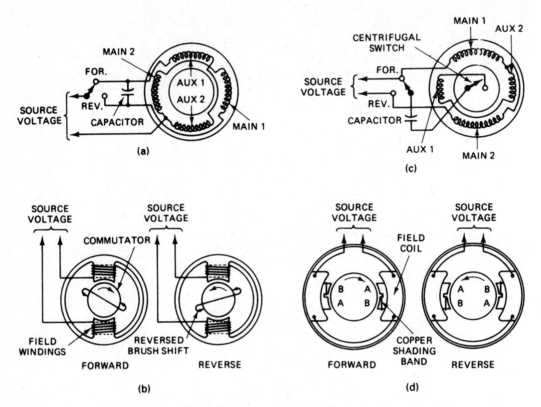

Figure 11–25. Techniques for controlling direction of rotation.

Ac motors that are brush equipped, as in drawing b, can have armature rotation changed by shifting brushes. This cannot be done with brushes that are fixed-position types, but some single-phase series motors have adjustable brush holders. The brush holder may be marked with arrows to indicate direction and may also have limit stops to show maximum brush-positioning stops.

When the motor reversing switch is in the forward position (drawing c), the phase shift capacitor is put in series with main winding 1. When the switch is in the reverse position, the capacitor is in series with main winding 2.

In drawing d, a change of direction can be had by transposing the shading rings.

Chapter 12

Solved Problems

Electrical problems are often encountered by electricians in their daily work, even though electrical repair and installation can become routine. However, for typical municipalities, advancement to the position of master electrician or passing a civil service examination for various levels requires some amount of problem solving.

CIRCULAR MILS

A circular wire having a diameter of 0.142 inch is used as a connecting wire to the motor in a washing machine, but the wire must be replaced. What is the minimum gauge of wire that should be used?

$$\text{Mils} = \text{inches} \times 1000 = 0.142 \times 1000 = 142 \text{ mils diameter}$$

$$\text{circular mils} = D \times D = 142 \times 142 = 20,164 \text{ circular mils}$$

The American Wire Gauge Table (see Table 3–3) shows that the nearest wire to meet these specifications is wire with gauge No. 7. This wire will have a slightly greater current-carrying capacity than the original.

Voltage Drop in a Feeder Line

In a factory an electric welder requires an operating current of 5 amperes, with a minimum input of 107 volts. The measured line voltage is 110 volts. The welder is to be operated at a distance of 400 feet from the point at which the power line enters the building. What size wire should be used to connect the input source and the welder?

The permitted voltage drop is $110 - 107 = 3$ volts

Table 3–3 in Chapter 3, shows that the wire most nearly matching this specification is No. 8 wire which has a circular mil (CM) area of 16,500.

$$\text{Circular Mils} = \frac{d \times I \times 22}{IR \text{ drop}}$$

d is the distance in feet, I is the current, and the denominator shows the voltage drop along the line.

$$\text{CM} = \frac{400 \times 5 \times 22}{3} = \frac{44,000}{3} = 14,666 \text{ CM}$$

RESISTOR CIRCUITS

Resistors can be used alone, in series, in parallel, or in series-parallel and can consist of two or more. In a series arrangement the total resistance is the sum of the individual resistors. The current flowing through the resistors is the same for all. The total resistance is the same regardless of the arrangement of the resistors.

Resistors in Series

$$R_t = R_1 + R_2 + R_3$$

What is the total resistance of three resistors having values of 20, 40, and 60 ohms wired in series?

$$R_t = 20 + 40 + 60 = 120 \text{ ohms}$$

Note that the formula does not take the tolerance of each resistor into consideration, and so the answer supplied by the formula is an approximate value.

Two Resistors in Parallel

The total resistance of resistors in parallel is always less than that of the resistor having the lowest ohmic value.

$$R_t = \frac{R_1 \times R_2}{R_1 + R_2}$$

R_t is the total resistance. For either series or parallel circuits, the values of the resistors must first be converted to similar units (ohms, kilohms, or megohms).

Two resistors having values of 80 and 37 ohms are wired in parallel. What is the total resistance?

$$R_t = \frac{80 \times 37}{80 + 37} = \frac{2960}{117} = 25.30 \text{ ohms}$$

Three Resistors in Parallel

The total resistance of three resistors in parallel can be calculated from:

$$R_t = \frac{1}{\dfrac{1}{R_1} + \dfrac{1}{R_2} + \dfrac{1}{R_3}}$$

What is the equivalent resistance of three resistors having values of 30, 45, and 63 ohms when wired in parallel?

$$R_t = \frac{1}{\dfrac{1}{30} + \dfrac{1}{45} + \dfrac{1}{63}} = \frac{1}{0.033 + 0.0222 + 0.0159} = \frac{1}{0.0711} = 14 \text{ ohms}$$

Resistors in Series-Parallel

The total resistance can be calculated by combining the formulas for series and parallel resistance.

$$R_t = R_1 + R_2 + \frac{1}{\dfrac{1}{R_1} + \dfrac{1}{R_2} + \dfrac{1}{R_3}}$$

RESISTANCE OF A LONG, COPPER WIRE

Length and Cross-Sectional Area

The resistance of a conductor is directly proportional to its length and inversely proportional to its cross-sectional area. In plain terms this statement means that the longer a wire, the greater its resistance, and the greater the volume of the wire, the lower will be its resistance. Stated in a formula:

$$R = \rho \frac{1}{a} \text{ (at 20°C)}$$

In this formula, R is the resistance of the conductor, l is its length in feet, and a is the cross-sectional area in circular mils. The letter ρ is the Greek letter rho and is used to denote the specific resistance.

As an example, what is the resistance of 2,000 feet of No. 18 copper wire at 20°C? In Table 3–3 in Chapter 3, locate No. 18 under the heading of Gauge Number. According to the chart, this wire has a cross section of 1,620 circular mils. You can locate this value by moving directly to the right from gauge No. 18, until you reach the column headed "circular mils." The specific resistance of copper wire at 20°C is 10.37, the value for ρ in the formula. We now have:

$$R = 10.37 \left(\frac{2,000}{1,620} \right) = 10.37 \times 1.235 = 12.80 \text{ ohms}$$

Go back to the chart and locate the column headed "25°C." This column supplies the resistance of 1,000 feet of No. 18 copper wire at 25°C. The value given is 6.51 ohms. For 2,000 feet, the resistance would be twice this amount or $6.51 \times 2 = 13.02$ ohms. Compare this resistance with the value obtained in the sample problem. It is 13.02 ohms for 2,000 feet of wire at 25°C versus 12.80 ohms for the same length of wire at 20°C. The difference in results is due to the difference in temperature. The same wire at 65°C, according to the chart, has a resistance of 7.51 ohms per thousand feet. For 2,000 feet of No. 18 wire, the resistance would be $7.51 \times 2 = 15.02$ ohms.

OHM'S LAW

Ohm's law for dc is available in three forms:

$$E = I \times R$$

$$I = \frac{E}{R}$$

$$R = \frac{E}{I}$$

In solving problems in a dc circuit, it is possible that one or more of these formulas will be required. In these formulas the voltage E is in volts, the current I is in amperes, and the resistance R is in ohms. Convert to basic units (Figure 12–1).

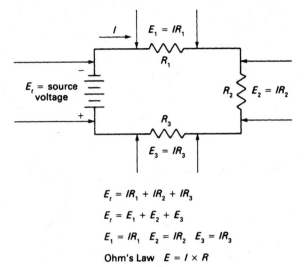

$$E_t = IR_1 + IR_2 + IR_3$$
$$E_t = E_1 + E_2 + E_3$$
$$E_1 = IR_1 \quad E_2 = IR_2 \quad E_3 = IR_3$$

Ohm's Law $\quad E = I \times R$

Figure 12–1. Applications of Ohm's law.

Voltage

A current of 650 milliamperes flows through a 10-ohm resistor. How much voltage appears across this resistor?

$$I = 650 \text{ milliamperes} = \tfrac{650}{1000} \text{ amperes} = 0.65 \text{ ampere}$$

$$R = 10 \text{ ohms}$$

$$E = 0.65 \times 10 = 6.5 \text{ volts}$$

Current

120 volts is measured across a 15-ohm resistor. How much current is flowing through the resistor?

$$E = 120 \text{ volts}$$

$$R = 15 \text{ ohms}$$

$$I = \frac{E}{R} = \frac{120}{15} = 8 \text{ amperes}$$

Resistance

What is the resistance of the heating element of a soldering iron when it is connected across a dc source voltage of 60 volts and a current of 4 amperes flows through it?

$$E = 60 \ volts$$

$$I = 4 \ \text{amperes}$$

$$R = \frac{E}{I} = \frac{60}{4} = 15 \ \text{ohms}$$

Voltage Divider

Three resistors, R_1, R_2, and R_3, (Figure 12–2) are series connected across a dc voltage source and are being used as voltage dividers to supply different voltages to a voltage distribution panel. R_1 is 18 ohms, R_2 is 33 ohms, and R_3 is 41 ohms. The current is 6 amperes. What is the voltage drop across each resistor? What is the total voltage? What is the total resistance in this circuit?

Since this is a series circuit, the same current flows through each resistor. The voltage across R_1 is based on

$$E = I \times R_1$$

$$= 6 \times 18 = 108 \ \text{volts}$$

The voltage across R_2 is calculated in the same way.

$$E = 6 \times 33 = 198 \ \text{volts}$$

And the voltage across R_3 is

$$E = 6 \times 41 = 246 \ \text{volts}$$

The total voltage (E_t) is the sum of the three voltages.

$$E_t = 108 + 198 + 246 = 552 \ \text{volts}$$

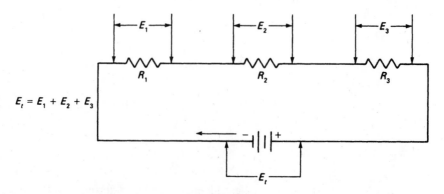

Figure 12–2. Voltage drops (IR drops) across series resistors.

The total resistance is the sum of the individual resistors.

$$\text{total resistance} = 18 + 33 + 41 = 92 \text{ ohms}$$

As a check on the work, find the total voltage in the circuit by multiplying the current by the total resistance. $E = 6 \times 92 = 552$ volts.

IR Drop

A current of 4 amperes flows through a resistor that has a value of 3 ohms. What is the *IR* drop (the voltage drop) across the resistor?

$$E = I \times R$$
$$= 4 \times 3 = 12 \text{ volts}$$

Branch and Line Currents

The current supplied by a power source, such as a dc generator or batteries, can branch off into a number of different circuit elements. The main current is called the line current. The sum of the branch currents is equal to the line current. Thus, $I_t = I_1 + I_2 + I_3 \ldots I_t$ is the line current, while I_1, I_2, and I_3 are three branch currents. There can be more or fewer branch currents, dictated by circuit requirements (Figure 12–3).

Voltage Drops (*IR* Drop)

Three lamps are connected in series across a voltage source. The lamps have hot resistances of 18, 21, and 34 ohms, respectively. The operating current is 3 amperes. How much voltage appears across each lamp? What is the total amount of voltage supplied?

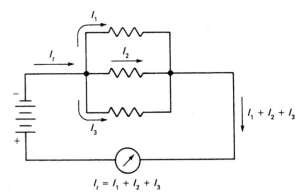

Figure 12–3. The line current (I_t) is equal to the sum of the branch currents.

$$E_t = IR_1 + IR_2 = IR_3 = E_1 + E_2 + E_3$$
$$E_t = (3)(18) + (3)(21) + (3)(34)$$
$$= 54 + 63 \text{ volts} + 102 \text{ volts}$$
$$= 219 \text{ volts}$$

Power in DC Circuits

Various formulas can be used for determining the power in a dc circuit. These are
$$W = E \times I$$
$$W = I^2 R$$
$$W = \frac{E^2}{R}$$

In electrical work, power, W or P, is generally expressed in watts or kilowatts. A kilowatt is 1000 watts or a watt is equal to $\frac{1}{1000}$ kilowatt.

When a lamp is connected to a dc source of 40 volts, a current of 2 amperes flows through it. How much power is being supplied?

$$E = 40, \qquad I = 2$$
$$W = E \times I$$
$$= 40 \times 2 = 80 \text{ watts}$$

A current of 65 milliamperes flows through a 15-ohm resistor. What is the power dissipation of this resistor?

$$I = 65 \text{ milliamperes} = \frac{65}{1000} = 0.065 \text{ ampere}$$
$$I^2 = I \times I$$
$$R = 15 \text{ ohms}$$
$$W = I^2 R = 0.065 \times 0.065 \times 15 = 0.063 \text{ watt}$$

Capacitors in Parallel

One problem in a relay circuit is insufficient capacitance. An existing capacitor has a value of 10 microfarads. Three additional capacitors are placed in parallel with it, and these have values of 6, 22, and 34 microfarads. What is the total capacitance of the four capacitors?

$$C_t = C_1 + C_2 + C_3 + C_4$$
$$= 10 + 6 + 22 + 34 = 72 \text{ } \mu F$$

Voltages in Series Aiding

DC generators or batteries are often wired in such a way that the voltages are additive. Such an arrangement is referred to as *series aiding*.

Four wet cells are wired in series aiding to supply the operating voltage required by a small dc motor. E_1 and E_2 are each 12 volts, while E_3 and E_4 are each 6 volts. What is the total available voltage?

$$E_t = E_1 + E_2 + E_3 + E_4$$
$$= 12 + 12 + 6 + 6 = 36 \text{ volts}$$

Coils in Series

Two coils are wired in series. The first has an inductance of 450 millihenrys, the second an inductance of 385 millihenrys. The coils are widely separated, so there is no magnetic coupling. What is the total inductance?

Inductance is represented by the letter L.

$$L_t = L_1 + L_2$$
$$= 450 + 385 = 835 \text{ mH}$$

Resistor Voltage Drops

A resistor is connected in series with a battery charger to limit the voltage being applied to a number of series-connected batteries. The current, I, is 500 milliamperes, and the resistor has a value of 3 ohms. What is the voltage across the resistor?

$$500 \text{ milliamperes} = \frac{500}{1000} \text{ ampere} = 0.5 \text{ ampere}$$
$$E = I \times R = 0.5 \times 3 = 1.5 \text{ volts}$$

A current of 50 milliamperes flows in a series circuit consisting of these resistors. R_1 is 40 ohms, R_2 is 20 ohms, and R_3 is 50 ohms. What is the voltage drop (*IR* drop) across each resistor? What is the source voltage?

$$\text{Ohm's law: } E = I \times R$$
$$E_1 = I \times R_1, \qquad E_2 = I \times R_2, \qquad E_3 = I \times R_3$$
$$E_t = IR_1 + IR_2 + IR_3$$

The current must be converted from milliamperes to amperes. To do this divide milliamperes by 1000 or move the decimal point three places to the left.

$$\frac{50\ \text{mA}}{1000} = 0.050\ \text{ampere} = 0.05\ \text{ampere}$$

$$E_1 = 0.05 \times 40 = 2\ \text{volts}$$

$$E_2 = 0.05 \times 20 = 1\ \text{volt}$$

$$E_3 = 0.05 \times 50 = 2.5\ \text{volts}$$

$$E_t = E_1 + E_2 + E_3 = 2 + 1 + 2.5 = 5.5\ \text{volts}$$

Tolerance

What are the upper and lower resistance limits of a resistor having a value of 500 ohms and a plus-minus tolerance of 10% (Figure 12–4).

$$\text{Tolerance} = 500 + 10\% = 500 + 50 = 550\ \text{ohms}$$

$$\text{Tolerance} = 500 - 10\% = 500 - 50 = 450\ \text{ohms}$$

$$\text{Tolerance} = 450\ \text{ohms to } 550\ \text{ohms}$$

DC Power

The voltage across a resistor in a dc circuit is measured at 43 volts. At that time, a current of 1450 milliamperes flows through the resistor. How much power is dissipated by the resistor?

The current of 1450 milliamperes must first be converted to amperes. To do so, divide the current by 1000.

$$\frac{1450}{1000} = 1.45\ \text{amperes}$$

$$W = E \times I$$

$$= 43 \times 1.45 = 62.35\ \text{watts}$$

A current of 95 milliamperes flows through a 65-ohm resistor. How much power does the resistor dissipate?

$$W = I^2 R$$

Convert 95 milliamperes to amperes by dividing it by 1000.

$$\frac{95}{1000} = 0.095\ \text{ampere}$$

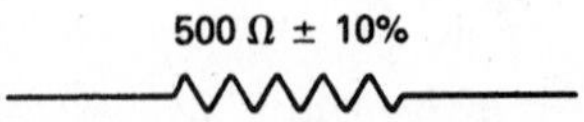

500 Ω ± 10%

Figure 12–4. Resistor tolerance.

$$I^2 = 0.095 \times 0.095 = 0.009$$
$$0.009 \times 65 = 0.585 \text{ watt}$$

A dc potential of 38 volts is measured across a 65-ohm resistor. How much power is being supplied to the resistor?

$$W = \frac{E^2}{R}$$

$$E^2 = E \times E$$

$$E^2 = 38 \times 38 = 1444,$$

$$R = 65$$

$$W = \frac{1444}{65} = 22.2 \text{ watts}$$

DC FORMULA MEMORY CIRCLE

The various forms of Ohm's law and the power laws for dc can be conveniently arranged in a circle, as in Figure 12–5. These formulas are for calculating power, current, voltage and resistance and are used for determining an unknown value when two values in a formula are known. The wheel can be used with 12 different formulas. Power is variously expressed as *P* or *W*.

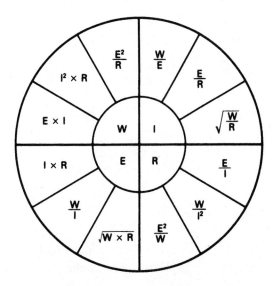

Figure 12–5.   Memory aid for dc formulas.

Temperature versus Wire Resistance

As an example, consider a wire whose resistance is 1 ohm at 0°C. If the temperature is increased to 10°C, the final resistance is:

$$R_t = 1 + (1)(10)(0.00427)$$

$$R_t = 1 + 0.0427 = 1.0427 \text{ ohms}$$

This is for a wire whose initial resistance is 1 ohm. For a wire whose resistance is 30 ohms at 0°C, with the same temperature increase of 10°:

$$R_t = 30 \ (30)(10)(0.00427)$$

$$R_t = 30 + 300 \times 0.00427$$

$$R_t = 30 + 1.281 = 31.281 \text{ ohms}$$

The problem shown above involves the temperature coefficient of resistance, but you can now use a formula such as:

$$R_t = R_o(1 + \alpha T)$$

R_t is the final resistance of the wire while R_o is the resistance of that same wire at 0°C. The letter α represents the temperature coefficient of resistance of copper wire at 0°C, or 0.00427. The letter α is the final temperature in degrees Celsius. While the formula may look difficult, you've been using it if you solved the earlier problem involving the resistance rise of a heated wire.

Consider a wire whose resistance is 30 ohms at 0°C, but whose temperature is increased to 10°C. Using the formula, substitute values given for the letters. R_o is 30; α is 0.00427; T is 10. This means, then, that $R_t = R_o(1 + \alpha T)$ will look like this:

$$R_t = 30(1 + 0.00427 \times 10) = 30(1 + 0.0427)$$

$$= 30(1.0427) = 31.281 \text{ ohms}$$

TEMPERATURE CONVERSIONS

In electrical work temperature is measured in either degrees Fahrenheit or Celsius. It is necessary to be able to convert from one to the other.

Degrees Celsius to Degrees Fahrenheit

After running for several hours, the armature of a motor had a measured temperature of 60°C. What was the temperature in degrees Fahrenheit?

$$F° = (C° \times \tfrac{9}{5}) + 32$$

$$F° = (60 \times \tfrac{9}{5}) + 32$$

$$F° = (\tfrac{60}{1} \times \tfrac{9}{5}) + 32$$

$$= 108°$$

Degrees Fahrenheit to Degrees Celsius

A bimetallic thermostat is designed to open when the temperature reaches 65°C. A thermometer adjacent to the thermostat reads 120°F. Will the thermostat open under this temperature condition?

$$C° = (F° - 32) \times \tfrac{5}{9}$$

$$C° = (120 - 32) \times \tfrac{5}{9} = 88 \times \tfrac{5}{9} = 48.89°C$$

The thermostat will not open.

Temperature Coefficient of Resistance

A certain length of No. 12 AWG bare copper wire has a resistance of 22 ohms at 0°C. What is its resistance at 77°F?

$$C° = (F° - 32) \times \tfrac{5}{9}$$

$$C° = (77 - 32) \times \tfrac{5}{9} = 45 \times \tfrac{5}{9} = 25°C$$

The temperature increase is from 0°C to 25°C.

$$R_t = R_o(1 + \alpha T) = 22 \, (1 + 0.00427 \times 25) = 25.107 \text{ ohms}$$

$$= 22 \times 1.10675 = 24.3485 \text{ ohms}$$

Watt-hours

An electric light bulb rated at 60 watts is turned on for 20 hours. How much electrical energy is supplied to the bulb?

$$\text{watt-hours} = \text{watts} \times \text{hours}$$

$$60 \times 20 = 1200 \text{ watt-hours} = 1.2 \text{ kilowatt-hours}$$

Power Usage

An electric range is rated at 4.8 kilowatts, while a broiler and toaster together require 1.4 kilowatts. What is the total power consumption in watts?

$$\text{watts} = \text{kilowatts} \times 1000$$
$$= (4.8 + 1.4) \times 1000$$
$$= (6.2 \times 1000 = 6200 \text{ watts}$$

Capacitors in Series

$$C_t = \frac{C_1 \times C_2}{C_1 + C_2}$$

Capacitors in Parallel

To increase the capacitance in a time-delay circuit, three capacitors are wired in parallel. Two capacitors are labeled 20 and 40 microfarads and the third was measured as 39 microfarads. What is the total capacitance?

$$C_t = C_1 + C_2 + C_3$$
$$C_t = 20 + 40 + 39 = 99 \text{ microfarads}$$

Voltages in Series

To obtain the voltage required to operate a small dc motor, three batteries are wired in series. The batteries are rated at 18, 22, and 88 volts. What is the total available voltage?

$$E_t = E_1 + E_2 + E_3$$
$$= 18 + 22 + 88 = 128 \text{ volts}$$

Coils in Series

Three iron-core coils used in the output of a dc power source to eliminate ripple and improve regulation have inductances of 8, 45, and 77 henrys, respectively. What is the total inductance? (Figure 12–6) There is no magnetic coupling between the coils.

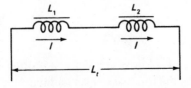

Figure 12–6. Coils (inductors) in series. There is no magnetic coupling between the coils.

$$L_t = L_1 + L_2 + L_3$$
$$= 8 + 45 + 77$$
$$= 130 \text{ henrys}$$

Coils in Series Aiding

When a pair of coils have interacting magnetic fields, the overall inductance can be increased or decreased by the effect known as mutual inductance (M).

A line noise filter consists of two coils, L_1 and L_2, wired in series aiding (aiding mutual inductance). The current flows through both coils in the same direction, thus producing aiding magnetic fields. L_1 is 465 millihenrys, L_2 is 456 millihenrys, and the mutual inductance is 67 millihenrys. What is the total inductance?

$$L_t = L_1 + L_2 + 2M$$
$$= 465 + 456 + 2 \times 67$$
$$= 465 + 456 + 134$$
$$= 1055 \text{ millihenrys}$$

Coils in Series Opposing

To prevent transient noise pulses traveling along a power line from triggering a relay unnecessarily, a pair of series coils is inserted in the line. The coil connections are arranged so that the coil currents are in opposite directions, with the result that the magnetic fields of the coils oppose each other. If L_1 has an inductance of 6 henrys, L_2 an inductance of 14 henrys, and the mutual inductance, M, is 2 henrys, what is the total inductance of the two coils? (Figure 12–7)

$$L_t = L_1 + L_2 - 2M$$
$$= 6 + 14 - (2 \times 2)$$
$$= (6 + 14) - 4$$
$$= 20 - 4 = 16 \text{ henrys}$$

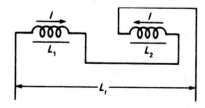

Figure 12–7. Coils with their magnetic fields in opposition.

SINE-WAVE VOLTAGES AND CURRENTS

These problems involve nothing more than knowing the arithmetic relationships between peak-to-peak, peak, rms, and average values of sine waves. The frequency of these waves, usually 60 hertz, is not involved.

Average Value of Voltage or Current Sine Wave

A sine-wave current flows through a noninductive resistor. The peak voltage measured across this resistor is 18.57 volts. What is the average voltage of this wave?

$$E_{av} = 0.637 \times E_{peak}$$
$$= 0.637 \times 18.57 = 11.829 \text{ volts}$$

The peak-to-peak voltage developed across a resistor is 63 volts. What is the average voltage for a single alternation of this sine wave?

A single cycle of a sine wave consists of two alternations, a positive half and a negative half. Each alternation is the mirror image of the other, and both have the same voltage values.

$$E_{av} = 0.3185 \times E_{p\text{-}p}$$
$$= 0.3185 \times 63 = 20.07 \text{ volts}$$

Peak Value of Voltage or Current Sine Wave

An average current of 385 milliamperes flows through a small lamp. The hot resistance of the lamp is 40 ohms. What is the peak voltage across the lamp? The peak current through the lamp? In amperes? In milliamperes?

Whenever solving electrical problems, it is necessary to work with basic units. As a first step, convert milliamperes to amperes. 385 milliamperes $= \frac{385}{1000}$ amperes $= 0.385$ ampere. Note that the problem specifies the hot resistance of the lamp. The cold resistance of a lamp is the minimum resistance of its filament. As the filament is heated, its resistance increases until it gradually reaches a steady amount. The filament has a positive temperature coefficient of resistance, meaning that its resistance increases with temperature.

$$385 \text{ milliamperes} = 0.385 \text{ ampere}$$
$$E_{av} = I_{av} \times R = 0.385 \times 40 = 15.4$$
$$E_{peak} = \frac{E_{av}}{0.637} = \frac{15.4}{0.637} = 24.18 \text{ volts}$$

$$I_{peak} = \frac{E_{peak}}{R} = \frac{24.18}{40} = 0.604 \text{ ampere}$$

$$0.604 \times 1000 = 605 \text{ milliamperes}$$

In a factory manufacturing electric toasters, completed units are selected at random and subjected to a variety of voltage tests. The average voltage measured across one such toaster is 95 volts. What is the peak voltage at which this toaster is being checked?

$$E_{peak} = 1.57 \times E_{av}$$

$$= 1.57 \times 95 = 149 \text{ volts}$$

The peak current can be calculated this way:

$$I_{peak} = \frac{I_{av}}{0.637}$$

$$= \frac{385}{0.637} = 604 \text{ milliamperes}$$

Average AC Power in a Resistive Circuit

A 1.2-kilohm resistor carries an average value of 120 milliamperes. What is the average power dissipated by this resistor?

$$P_{av} = I^2_{av} \times R$$

$$I = 120 \text{ milliamperes} = 0.12 \text{ ampere}$$

$$I^2 = 0.12 \times 0.12 = 0.0144$$

$$1.2 \text{ kilohms} = 1.2 \times 1000 = 1200 \text{ ohms}$$

$$P_{av} = 0.0144 \times 1200 = 17.28 \text{ watts}$$

RMS Value of Voltage or Current Sine Wave

A peak-reading ammeter indicates a current flow of 314 milliamperes. What is the rms value of this current?

No conversion to basic amperes is required in this problem since other electrical values are not used. The rms value is known as the effective value. If this peak current of 314 milliamperes were sent through a resistor, how much direct current would be needed to produce an equivalent heating effect? A current of 222 milliamperes dc would be required, corresponding to an effective current of 222 milliamperes ac.

$$I_{eff} = 0.707 \times I_{peak} = 0.707 \times 314 = 222 \text{ milliamperes}$$

Converting Peak-to-Peak Voltage to RMS

The voltage rating of an electrical component is indicated by the manufacturer as 300 volts, peak-to-peak. Would it be safe to connect this component across a voltage source whose output is indicated as 160 volts rms?

$$E_{eff} = \frac{E_{p-p}}{2.828}$$

$$= \frac{300}{2.828} = 106 \text{ volts}$$

The equivalent rms rating of the component is 106 volts. It would not be safe to connect it across 160 volts rms.

Peak-to-Peak Value of Voltage or Current Sine Wave

A peak-reading ac voltmeter measures 63.2 volts across a purely resistive element in a circuit. What is the peak-to-peak (p-p) voltage?

$$E_{p-p} = 2 \times E_{peak}$$

$$= 2 \times 63.2 = 126.4 \text{ volts}$$

Peak Power

Two appliances are operated in parallel across a 5-kilowatt generator supplying 110 volts peak. Their combined resistance is 10 ohms. Will the generator be overloaded if all these appliances are used at the same time?

$$P_{peak} = \frac{E_{peak}^2}{R}$$

$$E = 110 \text{ volts}, \qquad E^2 = 110 \times 110 = 12,100$$

$$P_{peak} = \frac{12,100}{10} = 1210 \text{ watts} = 1.2 \text{ kilowatts}$$

Since the peak power load is only 1.2 kilowatts, the generator will not be overloaded.

Average Voltage

The peak-to-peak voltage developed across a resistor is 63 volts. What is the average voltage for a single alternation of a sine wave?

$$E_{av} = 0.3185 \times E_{p\text{-}p}$$
$$= 0.3185 \times 63 = 20.07 \text{ volts}$$

Another method of solving this problem is to find the peak voltage and then multiply by 0.637.

$$E_{peak} = \frac{E_{p\text{-}p}}{2} = \frac{63}{2} = 31.5 \text{ volts}$$
$$E_{av} = 0.637 \times 31.5 = 20.07 \text{ volts}$$

A resistive element coated with a ceramic material is being used to heat liquids. The peak-to-peak voltage across this element is 120 volts. What is the average voltage?

$$\text{Peak-to-peak voltage} = 120$$
$$\text{Average voltage} = 0.3185 \times \text{peak-to-peak voltage}$$
$$= 0.3185 \times 120 = 38.22 \text{ volts}$$

Peak Voltage

The peak-to-peak voltage of a sine wave is measured at 90 volts. What is the peak voltage?

$$E_{peak} = \frac{E_{p\text{-}p}}{2} = \frac{90}{2} = 45 \text{ volts}$$

AC Voltage across a Capacitor

A capacitor having a reactance of 70 ohms is in a series circuit whose current is measured at 100 milliamperes. What is the voltage across the capacitor?

$$.00 \text{ milliamperes} = \frac{100}{1000} = 0.1 \text{ ampere}$$
$$E_c = I \times X_c = 0.1 \times 70 = 7 \text{ volts}$$

Applied voltage

In a series ac circuit, the voltage measured across the resistor is 60 volts, while that across an inductor is 90 volts. What is the amount of the source voltage?

$$E_{applied} = \sqrt{E_R^2 + E_L^2}$$
$$E_R = 60, \qquad E_R^2 = 60 \times 60 = 3600$$

$$E_L = 90, \qquad E_L^2 = 90 \times 90 = 8100$$
$$E_{applied} = \sqrt{3600 + 8100} = \sqrt{11,700} = 108 \text{ volts}$$

Applied Voltage in an RL Circuit

A current of 900 milliamperes flows in a series RL circuit supplied by a generator whose output frequency is 25 hertz. The coil has an inductance of 800 millihenrys and the resistor has a value of 50 ohms. What is the output voltage supplied by the generator to this circuit?

$$X_L = 6.28 \times f \times L = 6.28 \times 25 \times 0.8 = 126 \text{ ohms}$$
$$E = I \times Z \qquad Z = \sqrt{R^2 + X_L^2}$$
$$E = I \times \sqrt{R^2 + X_L^2}$$
$$R = 50, \qquad R^2 = 50 \times 50 = 2500$$
$$X_L = 126, \qquad X_L^2 = 126 \times 126 = 15,876$$
$$I = 900 \text{ milliamperes} = 0.9 \text{ ampere}$$
$$E = 0.9 \times \sqrt{2,500 + 15,876}$$
$$= 0.9 \times \sqrt{18,376} = 0.9 \times 135 = 121.5 \text{ volts}$$

Inductive Reactance

A coil used in a motor starter relay has an inductance of 300 millihenrys and is designed to work at 60 hertz. What is the reactance of this coil?

$$X_L = 6.28 \times f \times L$$
$$L = 300 \text{ millihenrys} = 0.3 \text{ henry}$$
$$X_L = 6.28 \times 60 \times 0.3 = 113 \text{ ohms}$$

AC Voltage across a Coil

Assume a current of 35 milliamperes flows through the coil described in the preceding problem. How much voltage develops across this coil?

$$E_L = I_L \times X_L$$
$$I_L = 35 \text{ milliamperes} = 0.035 \text{ ampere}$$
$$E_L = 0.035 \times 113 = 3.96$$

Current through a Coil

A normally open (NO) relay used across a 100-volt circuit works at a frequency of 60 hertz and has an inductance of 200 millihenrys. How much current flows through the coil?

$$X_L = 6.28 \times f \times L = 6.28 \times 60 \times .02 = 75 \text{ ohms}$$

$$I_L = \frac{E_L}{X_L} = \frac{100}{75} = 1.33 \text{ amperes}$$

Inductive Reactance

A voltage of 90 volts appears across a holding coil with a current of 410 milliamperes flowing through it. What is the reactance of the coil?

$$X_L = \frac{E_L}{I} = \frac{90}{0.41} = 219.5 \text{ ohms}$$

Capacitive Reactance

A 10-microfarad capacitor is used as a shunt across the 110-volt, 60-hertz power line. What is the reactance of this capacitor? What would be the reactance at half the frequency? At twice the frequency?

$$X_c = \frac{1}{6.28 \times f \times C} = \frac{1}{6.28 \times 60 \times 0.00001} = \frac{1}{0.00377}$$

$$X_c = 265 \text{ ohms}$$

$$2X_c = 265 \times 2 \text{ ohms} = 530 \text{ ohms}$$

$$X_c/2 = 265/2 = 132.5 \text{ ohms}$$

Impedance of an RL Circuit

A motor speed control unit includes a 100-ohm resistor (R) in series with a coil (L) whose reactance at the operating frequency is 60 ohms. What is the impedance of this R-L combination?

$$Z = \sqrt{R^2 + X_L^2}$$

$$R = 100 \text{ ohms}, \qquad R^2 = 100 \times 100 = 10,000$$

$$X_L = 60 \text{ ohms}, \qquad X_L^2 = 60 \times 60 = 3600$$

$$Z = \sqrt{10,000 + 3600} = \sqrt{13,600} = 116 \text{ ohms (approx.)}$$

Impedance of an RC Circuit

In a 60-hertz circuit, a capacitor (C) has a reactance of 40 ohms. This capacitor is wired in series with a 50-ohm resistor (R). What is the impedance of this R-C combination?

$$Z = \sqrt{R^2 + X_C^2}$$

$$R = 50 \text{ ohms}, \qquad R^2 = 50 \times 50 = 2500 \text{ ohms}$$

$$X_C = 40 \text{ ohms}, \qquad X_C^2 = 40 \times 40 = 1600 \text{ ohms}$$

$$Z = \sqrt{2500 + 1600} = \sqrt{4100} = 64 \text{ ohms}$$

Voltage

The current flowing in the circuit of the preceding problem is 300 milliamperes. How much voltage is applied to the R-C circuit?

$$I = 300 \text{ milliamperes} = 0.3 \text{ ampere}$$

$$E = I \times Z = 0.3 \times 64 = 19.2 \text{ volts}$$

Impedance

The voltage applied to a series RL circuit is 120 volts. Under these conditions, a current of 500 milliamperes flows through the resistor and the coil. What is the impedance of this circuit?

$$Z = \frac{E}{I}$$

$$500 \text{ milliamperes} = 0.5 \text{ ampere}$$

$$Z = \frac{120}{0.5} = 240 \text{ volts}$$

Current

How much current will flow through an impedance of 85 ohms when the circuit voltage across the impedance is 200?

$$I = \frac{E}{Z} = \frac{200}{85} = 2.35 \text{ amperes (approximately)}$$

Primary Current of a Transformer

A power transformer connected to a 110-volt line has a secondary current of 750 milliamperes and develops 700 volts across its secondary. What is the primary current?

$$I_1 = \frac{E_2 \times I_2}{E_1} = \frac{700 \times 0.75}{110}$$

$$= 4.8 \text{ amperes}$$

A transformer with an 8:1 turns ratio has a secondary current of 650 milliamperes. What is the amount of current flowing in the primary winding?

$$I_1 = \frac{I_2 \times N_2}{N_1}$$

$$= \frac{0.650 \times 8}{1} = 5.2 \text{ amperes}$$

If the transformer is connected to a 110-volt power line, what is the power input?

$$P_i = E_1 \times I_1 = 110 \times 5.2 = 572 \text{ watts}$$

What is the amount of secondary voltage?

$$E_2 = \frac{E_1 \times N_2}{N_1} = \frac{110 \times 8}{1} = 880 \text{ volts}$$

What is the amount of power output at 100% efficiency?

$$P_o = E_2 \times I_2 = 880 \times 0.650 = 572 \text{ watts}$$

What is the power output at 90% efficiency?

$$P_o = 572 \times 0.90 = 514.8 \text{ watts}$$

What is the power loss in the transformer?

$$\text{power loss} = 100\% - 90\% = 10\%$$

$$10\% \times \text{power output} = 0.10 \times 572 = 57.2 \text{ watts}$$

$$\text{power lost in the transformer} = 57.2 \text{ watts}$$

$$\text{power output across the secondary} = 514.8 \text{ watts}$$

$$\text{total power developed} = 572 \text{ watts}$$

Efficiency Factor

A motor generator requires an input of 600 watts when connected to a full load. The maximum power delivered by the generator side of the unit is 375 watts. What is the efficiency factor of this unit? (Figure 12–8)

$$\text{Efficiency factor} = \text{output/Input}$$

$$\text{Efficiency factor} = \frac{375}{600} = 0.625$$

Note that the efficiency factor is measured when the transformer is working under its rated load.

Efficiency and Horsepower

The efficiency factor can be combined with formulas involving either power in terms of watts, or horsepower. The formula for horsepower is:

$$\text{hp} = \frac{E \times I}{746}$$

Transposing:

$$746\,\text{hp} = E \times I$$

$$I = \frac{746 \times \text{hp}}{E}$$

If the efficiency is less than 100%, the formula becomes:

$$I = \frac{746 \times \text{hp}}{E \times \text{eff}}$$

How much current will be required by a 5-horsepower, 220-volt dc motor whose operating efficiency is 90%?

$$I = \frac{746 \times 5}{220 \times 0.9} = 18.84 \ \text{amperes}$$

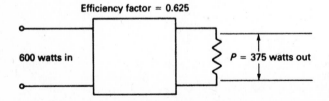

Efficiency factor = 0.625

600 watts in

P = 375 watts out

Figure 12–8. The efficiency factor is the ratio of the output to the input, in watts.

True Power

A circuit having a 75% power factor has a current of 600 milliamperes and an emf of 90 volts. What is the true power (real power) in this circuit?

As a first step, convert milliamperes to amperes.

$$600 \text{ milliamperes} = \frac{600}{1000} = 0.6 \text{ ampere}$$

Convert the power factor from a percentage to a decimal.

$$75\% = 0.75 \text{ by moving the decimal point}$$
$$\text{two places to the left}$$

$$\text{true power} = E \times I \times PF = 90 \times 0.6 \times 0.75$$

$$= 40.5 \text{ watts}$$

The voltage developed in a reactive circuit having a 65% power factor is 90 volts. The impedance in the circuit is 30 ohms. What is the true power?

$$P = \frac{E^2}{Z} \times PF$$

$$E = 90 \text{ volts}, \qquad E^2 = 90 \times 90 = 8,100$$

$$P = \frac{8100 \times 0.65}{30} = \frac{810 \times 0.65}{3} = 270 \times 0.65 = 175.5 \text{ watts}$$

A 500-milliampere current flows in a reactive circuit having an impedance of 20 ohms and a 90% power factor. What is the true power?

$$I = 500 \text{ milliamperes} = 0.5 \text{ ampere}, \qquad I^2 = 0.5 \times 0.5 = 0.25$$

$$P = I^2 \times Z \times PF$$

$$= 0.25 \times 20 \times 0.9 = 4.5 \text{ watts}$$

Peak AC Power in a Resistive Circuit

An rms current of 90 milliamperes flows through a 50-ohm resistor. What is the peak ac power dissipated by the resistor? The peak current is:

$$I_{peak} = \frac{I_{eff}}{0.707}$$

$$= \frac{90}{0.707} = 127 \text{ milliamperes}$$

The peak power is

$$P_{peak} = I^2_{peak} \times R$$

$$I_{peak} = 127 \text{ milliamperes} = 0.127 \text{ ampere}$$

$$I^2_{peak} = 0.127 \times 0.127 = 0.016 \text{ ampere}$$

$$P_{peak} = 0.016 \times 50 = 0.8 \text{ watt}$$

Average AC Power in a Resistive Circuit

The peak voltage developed across a 100-ohm resistor is 35 volts. What is the average power used by this resistor?

$$E_{av} = 0.637 \times E_{peak} = 0.637 \times 35 = 22.3 \text{ volts}$$

$$P_{av} = \frac{E^2_{av}}{R} = \frac{22.3 \times 22.3}{100} = \frac{497.29}{100} = 4.97 \text{ watts}$$

Current in a Reactive Circuit

A reactive circuit develops 50 watts with an applied voltage of 100 volts and a power factor of 80%. How much current flows in the circuit?

$$I = \frac{P}{E \times PF} = \frac{50}{100 \times 0.8} = \frac{50}{80} = \frac{5}{8} \text{ ampere}$$

A circuit having an impedance of 40 ohms and a 50% power factor is rated at 50 watts. How much current flows in this circuit?

$$I = \sqrt{\frac{P}{Z \times PF}}$$

$$= \sqrt{\frac{50}{40 \times 0.5}} = \sqrt{\frac{50}{20}} = \sqrt{2.5} = 1.6 \text{ amperes}$$

A power of 60 watts in a purely resistive ac circuit develops across a 10-ohm resistor. How much current flows in this circuit?

$$I = \sqrt{\frac{P}{R}} = \sqrt{\frac{60}{10}} = \sqrt{6} = 2.45 \text{ amperes}$$

Source Voltage

A resistor is wired in series with a coil. The measured voltage across the resistor is 37 volts and that across the coil is 11 volts. What is the amount of source voltage?

$$E_{source} = \sqrt{E_R^2 + E_X^2}$$

$$= \sqrt{37^2 + 11^2}$$

$$= \sqrt{1369 + 121}$$

$$= \sqrt{1490}$$

$$= 38.6 \text{ volts}$$

Power Factor

This problem involves finding the square root of a number. The easiest and fastest way of solving square root problems is to use a calculator.

An ac circuit includes a 40-ohm resistor and a 70-ohm reactance. What is the PF?

$$Z = \sqrt{R^2 + X^2}$$

$$R = 40 \text{ ohms}, \qquad R^2 = 40 \times 40 = 1600$$

$$X = 70 \text{ ohms}, \qquad X^2 = 70 \times 70 = 4900$$

$$Z = \sqrt{1600 + 4900} = \sqrt{6500} = 80 \text{ ohms (approximately)}$$

$$PF = \frac{R}{Z} = \frac{40}{80} = 0.5$$

$$0.5 \times 100 = 50\% \text{ power factor}$$

A reactive circuit develops 85 watts as measured by a wattmeter. An ammeter in the 65-volt circuit reads 2 amperes. What is the power factor?

$$PF = \frac{\text{watts}}{\text{volt-amperes}} = \frac{85}{2 \times 65} = \frac{85}{130} = 0.65$$

$$0.65 \times 100 = 65\% \text{ power factor}$$

Reactance

$$\text{Inductive:} \quad X_L = 2\pi fL$$

$$\text{Capacitive:} \quad X_C = \frac{1}{2\pi fC}$$

The letter X is used to indicate reactance, with X_L *for inductive reactance and X_C for capacitive reactance. Both types in ac circuits are measured in ohms. The letter π is equal to* 3.1416; f is the frequency in hertz (cycles per second), L the inductance in henrys, and C the capacitance in farads. If a submultiple is supplied in millihenrys or microhenrys, it should be converted to henrys before using in a formula. If the compacitance is supplied in microfarads or picofarads, change it to farads prior to using in a formula.

Impedance

$$Z = \frac{E}{I}$$

$$Z = \sqrt{R^2 + X^2}$$

$$Z = \sqrt{R^2 + (X_L - X_C)^2}$$

$$Z = \sqrt{R^2 + (X_C - X_L)^2}$$

Impedance, represented by the letter Z, is measured in ohms. Like inductive and capacitive reactance, it is used in ac circuits, but it also finds applications in circuits having a varying current. Two of the preceding formulas involve inductive and capacitive reactance. If the inductive reactance is greater than the capacitive reactance, it is shown as $X_L - X_C$. If the capacitive reactance is larger, it is listed as $X_C - X_L$.

Single-, Two-, and Three-phase Power Formulas

Single-phase circuits.

$$W = E \times I \times \text{PF}$$

$$W = I^2 \times R$$

$$I = \frac{W}{E \times \text{PF}}$$

$$E = \frac{W}{I \times \text{PF}}$$

$$PF = \frac{W}{E \times I}$$

Two-phase circuits.

$$W = E \times I \times PF \times 2$$

$$I = \frac{W}{E \times PF \times 2}$$

$$E = \frac{W}{I \times PF \times 2}$$

$$PF = \frac{W}{E \times I \times 2}$$

Three-phase circuits.

$$W = E \times I \times PF \times 1.732$$

$$I = \frac{W}{E \times PF \times 1.732}$$

$$E = \frac{W}{I \times PF \times 1.732}$$

$$PF = \frac{W}{E \times I \times 1.732}$$

W is the power in watts, E is the voltage in volts, I is the current in amperes, and PF is the power factor, which is usually expressed as a percentage; it must be converted to a decimal before being used in a formula. 1.732 is the square root of 3.

Single-phase Kilovolt-amperes

A single-phase ac generator has an output of 220 volts and delivers a maximum of 20 amperes to a load. What is the kVA rating of this generator?

$$kVA = \frac{E \times I}{1000}$$

$$kVA = \frac{220 \times 20}{1000} = \frac{4400}{1000} = 4.4 \ kVA$$

Three-phase Kilovolt-amperes

A three-phase ac generator has an output of 208 volts. With rated maximum load, the generator can deliver 40 amperes. What is the kVA rating of this generator?

$$kVA = \frac{E \times I \times 1.732}{1000} = \frac{208 \times 40 \times 1.732}{1000}$$

$$kVA = \frac{14,410}{1000} = 14.4 \ kVA$$

Power in a Single-phase Reactive Circuit

A 10-ampere load having a 60% power factor is connected to a 115-volt line. What is the real power used by this load?

$$P = E \times I \times PF = 115 \times 10 \times 0.6 = 1150 \times 0.6 = 690 \text{ watts}$$

Power in a Three-phase Reactive Circuit

How much power is used by a load on a three-phase circuit if the line voltage is 240 volts, the line current is 5 amperes, and the power factor is 80%?

$$P = E \times I \times PF \times 1.732$$

$$= 240 \times 5 \times 0.8 \times 1.732$$

$$= 1663 \text{ watts}$$

Sometimes the wording of a problem may seem to be so confusing that it might appear to require an unusual or different formula. Consider a problem such as this one: A three-phase, star-connected load having a reactive element is connected to a 240-volt circuit. It draws a current of 10 amperes and has a 50% power factor. How much power is used by the load?

The question supplies more information than required. Ignore "three-phase, star-connected, and reactive element." The useful information is the voltage, the current, and the power factor. The remainder is covered by the formula:

$$P = E \times I \times PF \times 1.732$$

$$P = 240 \times 10 \times 0.5 \times 1.732 = 2078 \text{ watts}$$

The answer is a reasonable approximation. The line voltage and current may be larger or smaller than that given in the data supplied. This is true not only in this problem but in others as well.

Current in a Three-phase Reactive Circuit

How much current will be taken by a 4-kilowatt load from a three-phase circuit whose line voltage is 440 volts at a 50% power factor?

$$4 \text{ kilowatts} = 4000 \text{ watts}$$

$$I = \frac{P}{E \times PF \times 1.732} = \frac{4000}{440 \times 0.5 \times 1.732}$$

$$= \frac{400}{44 \times 0.5 \times 1.732} = \frac{400}{22 \times 1.732} = 10.49 \text{ amperes}$$

Voltage in a Three-phase Reactive Circuit

A three-phase system operates with a 5000-watt load, drawing 10 amperes and working with a power factor of 50%. What is the voltage of the system?

$$E = \frac{P}{I \times PF \times 1.732}$$

$$= \frac{5000}{10 \times 0.5 \times 1.732} = \frac{5000}{5 \times 1.732} = \frac{1000}{1.732} = 577.37 \text{ volts}$$

Reading a Kilowatt-hour Meter

A kilowatt-hour meter has the following pointer settings. Reading from left to right: pointer between 3 and 4; pointer between 6 and 7; pointer directly opposite 2; pointer between 8 and 9. The previous reading was 3245 kilowatt-hours. The energy consumption for billing purposes is:

$$3628 - 3245 = 383 \text{ kilowatt hours}$$

If electrical energy is sold at the rate of 9¢ per kilowatt-hour the cost is 383×0.09 = $34.47.

Electrical Energy

In a home, six 40-watt light bulbs were turned on for 8 hours, a television set rated at 300 watts was on for 2 hours, while an air conditioner using 3,000 watts was on for 5 hours. What was the total use of electrical energy in kilowatt hours?

$$6 \times 40 = 240 \text{ watts} \quad 240 \times 8 = 1,920 \text{ watt-hours}$$
$$300 \times 2 = 600 \text{ watt-hours}$$
$$3,000 \times 5 = 15,000 \text{ watt-hours}$$
$$\overline{17,520 \text{ watt-hours}}$$
$$= 17.52 \text{ kilowatt-hours}$$

Horsepower Formulas

$$\text{dc:}\quad \text{hp} = \frac{E \times I \times \text{eff.}}{746}$$

$$\text{One-phase:}\quad \text{hp} = \frac{E \times I \times \text{eff.} \times \text{PF}}{746}$$

$$\text{Two-phase:}\quad \text{hp} = \frac{E \times I \times \text{eff.} \times \text{PF} \times 2}{746}$$

$$\text{Three-phase:}\quad \text{hp} = \frac{E \times I \times \text{eff.} \times \text{PF} \times 1.732}{746}$$

A motor is rated at 1.6 kilowatts. What is its equivalent horsepower (hp) rating?

$$1.6 \text{ kilowatts} = 1600 \text{ watts}$$

$$0.001341 \times W = \text{hp}$$

$$1600 \times 0.001341 = 2.15 \text{ hp}$$

A generator weighing 3 tons has been delivered to a power substation and is to be lifted through a distance of 20 feet. The crane that will lift the generator can do this in 2 minutes. How much horsepower is required? Assume that a ton consists of 2000 pounds.

$$3 \times 2000 = 6000 \text{ pounds.}$$

The number of foot-pounds per minute will be

$$\frac{6000 \times 20}{2}$$

Why $\frac{20}{2}$? If the crane can lift the generator 20 feet in 2 minutes, it can lift it 10 feet in 1 minute. By definition, horsepower is the weight lifted per minute.

$$\text{horsepower} = \frac{6000 \times 20}{2 \times 33,000} = 1.818$$

The wattage equivalent is $1.818 \times 746 = 1356$ (1 hp = 746 watts).

Amperes When Horsepower Is Known

$$\text{dc:}\quad I = \frac{746 \times \text{hp}}{E \times \text{eff.}}$$

$$\text{One-phase:}\quad I = \frac{746 \times \text{hp}}{E \times \text{eff.} \times \text{PF}}$$

$$\text{Two-phase:} \quad I = \frac{746 \times \text{hp}}{E \times \text{eff.} \times \text{PF} \times 2}$$

$$\text{Three-phase:} \quad I = \frac{746 \times \text{hp}}{E \times \text{eff.} \times \text{PF} \times 1.732}$$

One horsepower = 746 watts and is equal to 550 foot-pounds per second. This unit of mechanical power is also equivalent to raising 33,000 pounds vertically 1 foot in 1 minute. One watt = $\frac{1}{746}$ hp = 0.00134 hp. I is the current in amperes; E is the voltage in volts.

Amperes When Kilowatts Are Known

$$\text{dc:} \quad I = \frac{\text{kW} \times 1000}{E}$$

$$\text{One-phase:} \quad I = \frac{\text{kW} \times 1000}{E \times \text{PF}}$$

$$\text{Two-phase:} \quad I = \frac{\text{kW} \times 1000}{E \times \text{PF} \times 2}$$

$$\text{Three-phase:} \quad I = \frac{\text{kW} \times 1000}{E \times \text{PF} \times 1/732}$$

Volt-Amperes

For dc circuits involving resistance, power can be expressed in terms of either watts or volt-amperes (volts multiplied by amperes). Written as VA, it is also applicable in ac circuits when the voltage and current are in phase. Power factor (PF) is not involved in dc circuits. For ac circuits, VA is applicable when the power factor is 1 (100%).

Index